Roo Niermann
314-914-5???

GUNN**SIGHTS**

Taking Aim on Selling in the High-Stakes Industry of International Aerospace

GUNNSIGHTS

by **Tom Gunn**

NAVAL INSTITUTE PRESS
Annapolis, Maryland

Naval Institute Press
291 Wood Road
Annapolis, MD 21402

© 2008 by Thomas Gunn

All rights reserved. No part of this book may be reproduced or utilized in any form or by any means, electronic or mechanical, including photocopying and recording, or by any information storage and retrieval system, without permission in writing from the publisher.

Library of Congress Cataloging-in-Publication Data

Gunn, Thomas, 1943–
  Gunn sights : taking aim on selling in the high stakes industry of international aerospace / Tom Gunn.
      p. cm.
  Includes index.
  ISBN 978-1-59114-346-8 (alk. paper)
  1. Aerospace industries. 2. Aircraft industry. 3. Marketing. I. Title.
  HD9711.5A2G86 2008
  629.1068'8—dc22   2008023310

Printed in the United States of America on acid-free paper ∞

14 13 12 11 10 09 08 9 8 7 6 5 4 3 2
First printing

*For Kate, light of my life, and our children
Tim, Kevin, Maureen, and Megan—
all of whom may have wondered
from time to time
just what it was I did for a living.*

Contents

Preface | ix

Introduction | 1

Chapter One
The Business of Aerospace | 11

Chapter Two
People | 36

Chapter Three
The Business of Selling | 72

Chapter Four
Selected Case Studies | 108

Chapter Five
The Process | 160

Epilogue | 181

Index | 185

Preface

Let me guess: you've picked up my book, and you're wondering, does the world really need another how-to business memoir written by a retired aerospace executive?

I don't know about "the world," but if *you* are an aerospace executive, you might find this book to be of some value. If you are an aerospace executive at one of the large defense contractors that lost a competition to my company—McDonnell Douglas—you might learn how we did it. If you are a financial analyst or fund manager, you might discover some things worth exploring beyond a balance sheet and behind a 10K. If you are a student, or an intern, junior trainee, an up-and-coming manager or director at any large company, not just in aerospace or defense contracting, you might learn how to better serve your customers, your employer, and yourself.

This is not a book about engineering, research and development, process planning, manufacturing, cost containment, or human resources, but it does touch on those topics. It is a book about selling before, during, and after development of a product. It's about selecting and assembling a professional sales staff; developing a process to shape your effort and keep all members of your team headed in the same direction; strategic planning; assessing the competition and understanding the customer; and finally, it's about the proper (with some comment on the often improper) interrelation of politics, politicians, and lobbyists.

My credibility? Yes, I'm a retired aerospace executive who has been credited with organizing and leading the teams that in 22 years sold more than $250 billion worth of airplanes, helicopters, and missiles. But, why me? Why did I write this book? I've read many business books, some of which have been well-written but offer forgettable messages. Too many of them speak more about the ego of the writer than the needs of the reader. Most offered glittering platitudes and rules of business behavior, but lacked real-world relevance. Even among the best of them, few changed my way of doing things, only my way of perceiving things.

For a long time, I have wanted to pass on to younger marketers certain markers that will give them some preview of the road in front of them. Yes, the traditional path to wisdom may begin at the bottom with training and education and reading, but none of these prepare the neophyte for what happens at the top, enjoying the triumphs but surviving the mistakes, the really hard losses that you never forget but know they made you smarter and better prepared for life.

There is a big difference between the basics—suiting up for work, looking good, showing up for meetings on time—and being the warrior who can stand and fight in the international arena, who will be trusted to do the right thing, make the right moves, and win (most of the time). There is more to salesmanship than a smile on the face and a shine on the shoes.

Before we plunge ahead, I'd like to make a few administrative remarks and acknowledgments. This is not a history of McDonnell Douglas filled with documented facts, but rather a memoir, drawn from the collective recollections of myself, my family, and many of my former associates who came forward with ideas, suggestions, criticisms, and compliments. I couldn't have done my job then, nor written this book now, without the assistance of Mark Bass, Jim Caldwell, Chris Chadwick, Mike Coggins, Bob Craighead, Darryl Davis, John Feren, Martin Fisher, Al Gingrich, Bob Grace, George Hibbard, Steve Krause, Mark Kronenberg, Cindy Malawy, Larry Merritt, Gerry Olsen, Chris Raymond, Jim Restelli, George Roman, Regina Stroupe, Mark Sullivan, Len Tavernetti, Stu Thompson, John Van Gels, Rich Weinstock, and Fred Whiteford.

While there may be an emphasis on international sales, most comments and suggestions have applicability in both international and domestic (U.S. military) markets even if that is not spelled out in the text. Throughout, "the company" means, in general, McDonnell Douglas or some operating element thereof: MDC Corporation, or McDonnell Aircraft (fighters), Douglas Aircraft (airliners, freighters, airlifters, and tankers), the Helicopter Company, Astronautics (space and missile systems), and, after 1997, Boeing. "We" means the company, or a staff on which I worked, the sales team, or me and someone else. I trust the context should be clear. Since most of our customers and almost everyone I worked with were men, my general use of male pronouns is not sexist insensitivity.

And, before I forget my manners, I must also thank Brayton Harris, retired naval officer, former aerospace executive, and accomplished author in his own right, who provided invaluable advice and assistance. I started this project with a hodge-podge of memories, an eclectic but incomplete collection of memorabilia, and myriad suggestions from my associates. Brayton helped me put them all in order, so the book actually has a beginning, a middle, and an Epilogue.

Introduction

We are overwhelmed with offers for quick-fix self-help books or DVDs ("Only pay shipping!") that offer simple solutions to make us healthy and wealthy, simplify our lives, and make our lives more fun and more satisfying.

This book is not one of them. There is nothing "quick" or simple about selling billions of dollars worth of high-performance airplanes. In truth, there are no shortcuts to marketing success, no matter what your business, whether it is big or small, or you are selling products or services. You need to study the market and develop a business plan. Is there a demand for what you plan to offer? Is there enough room in the market for another competitor? Why would a buyer choose you over the others and what can you do to sharpen the discriminators (price, quality, delivery time)? Will the market be around for a long time or is this a short-term opportunity? Either way, can you charge enough to cover your investment and make money?

The market is constantly moving from yesterday to the future. You need to keep up—keep ahead—with good employees, research and development (R&D), partnerships, and acquisitions to ensure that your offering remains better, more efficient, more lethal (in our business), more attractive to the customer, and better suited to the mission.

Every new campaign should start with a review of your business plan. Is it still valid? Is there still a sufficient market, not just with this potential customer, or is it approaching the end, overtaken by technology or changing tactics. Perhaps it is not worth tying up resources to pinch off a few more sales. If there is a market, do you have the right resources dedicated or available? Have your competitors developed something new? Do you think the customer will see enough difference between your product and theirs?

As you read through those last two paragraphs, were you working the questions in your mind? Welcome to my world. Read on, and discover how I got there.

One of my grandfathers, newly arrived from Ireland, was a streetcar conductor, the other, a first-generation Irish-American, was a steam fitter. They both knew that education was the key to success in America and really forced it on their children. My father, Donald Gunn, paid his way through law school acting on the Chautauqua circuit (sort of an intellectual vaudeville). My mother, Loretto Hennelly, earned her way through college (the first woman in her family to get a degree) playing the piano and singing on the same circuit. In 1928, as members of a three-man two-woman team on a 16,000 mile swing through five Southern states, they fell in love. They married, settled in St. Louis, and in twenty-four years had nine children—five boys, four girls, with me in the middle. In about the same time, the family went from almost-poor, where my brother Mike and I shared the bottom bunk of a double-decker, to living in a house with twenty-eight rooms.

You might think that with this somewhat Bohemian background, our family life would be a bit laissez-faire. You would be overlooking the strict, old-school, Irish-Catholic ethos. We lived in a two-class system: the girls washed, cooked, and cleaned, and the boys did all of the heavy labor and a summer job was mandatory. When it came to education and career choices, Father Knew Best. The girls all earned a college degree and grad school was a plus. The plan for the boys was to become lawyers and join his firm. Sort of like interchangeable parts of the same family machine.

Everyone went along with the plan except me.

I didn't know what I wanted to do when I grew up, but I knew pretty early on that it would be something other than working for my father. I was, in truth, the alien, the black sheep of the family, an outsider in the midst. I chaffed at toeing any line. If I didn't have a daily fistfight with one of my brothers, I had one going to school or coming home from school.

INTRODUCTION

In Summer 1958, to improve my fighting skills, I began taking boxing lessons at the Falcon Club, a Polish-American hangout where I not only could box, but could smoke and sneak some beer. By Fall, I was pronounced good enough to enter a qualifying match for the Golden Gloves. That was a banner year for the Golden Gloves, when Kentucky's seventeen-year-old Cassius Clay and Missouri's fifteen-year-old Tom Gunn took the first stepping stone to professional ranks, and (for me) perhaps a ticket to something more interesting than the family business. Clay began winning Golden Gloves championships, turned pro, changed his name to Muhammad Ali, and entered the ranks of boxing greats. I learned a couple of life changing lessons.

I invited my friends to the fight—a handful of teenagers with attitude, each wearing a white T-shirt with a cigarette pack rolled up in the sleeve. The fight was scheduled as two rounds, with two minutes a round. Going in, I suspected that I might be in trouble when my opponent showed up with a silk robe and his name embroidered on the back. I *knew* I was in trouble when the referee said, "Show me your cups." No one could box in a Golden Gloves match without an athletic cup to protect the privates, as if a knee to the groin was a standard move. I wasn't wearing a cup.

The ref was about to declare the match for my opponent, by default, when I remembered, "I have a cup! It's in my gym bag, downstairs." It was a soccer cup, but is not a cup a cup? With the ref's grudging approval, I jumped down from the ring, ran through the crowd, down to the basement dressing room, grabbed the cup, and realized that I didn't have a jock strap to keep it in place. Nonetheless, I stuffed it in my trunks, ran all the way back upstairs holding my crotch, and, gasping for breath, presented myself "Ready!"

The bell rang and we started. I took the first punch. The cup fell out on the floor. The spectators burst out laughing, my friends the loudest among them. We paused, while I stuffed the cup back in my trunks. It happened again, several times, too many times. I can't remember exactly how many but enough for the two rounds to become an eternity of humiliation.

Well, I was a tough kid and absorbed the shame of the loss, but I took away two vital lessons that you will find repeated, often, in the pages that follow:

1. Always know the rules of the game before you decide to play.
2. Be prepared.

The punches that I took (in between grabbing my crotch) must have knocked some sense into my rebellious brain. Teachers had always cited me as disruptive and a poor scholar. I guess I was. From this point on, I tried harder, and made some progress. I won an acting contest that came with a scholarship to go to California to study acting, but my father said no. I won an academic scholarship to go to a school in Texas. My father said no. In college, my political science advisor wanted to nominate me as an intern at the World Court in The Hague. My father said no. Oh, how I resented what I saw as a power-play, his need to be in control. My plans didn't fit his plan. With mature hindsight—would that all sons could learn to understand their fathers—I can affirm: He knew what I was not ready to admit; that I was not yet ready to go off the reservation, away from the steadying influence of church and family.

So, yes, living at home and with a B.A. in Political Science in my pocket, I went to law school and became a lawyer (and got married to the love of my life, Kate, then and evermore). However, rather than join my father immediately as expected, I came up with a very convoluted argument that I should first go to Washington, D.C., get a government job, and gain solid political experience.

My family was neck-deep in politics. My father started with an appointment in the Truman administration, as collector of revenue at the St. Louis regional IRS (proffered by Harry Truman, personally, on the same day the president fired Gen. Douglas MacArthur). My father became president of the St. Louis Board of Aldermen and with two other Irish-Americans ran the local Democratic Party. The children of the recently-arrived Irish knew that politics gave power, and that power controlled jobs, which in turn provided support to politicians. My mother had been chairperson for Stuart Symington's successful senatorial campaigns. My cousin, Tom Eagleton, was Missouri's lieutenant governor and gearing up for a run for the senate. In my view, my father's law practice certainly would benefit from having an experienced, politically-savvy son on the payroll. My father saw through that argument. In his eyes, I was thenceforth, and for a long time, the family traitor.

My mother, as mothers often are, was more understanding. A job with the government might be just the thing to steady me down. She suggested that I get in touch with Sen. Symington, the senator from Missouri, which I did, and I took a job as an attorney-examiner with the Federal Trade Commission. My mother's brother, Mark Hennelly, was very close to Arkansas Sen. John McClellan and set up a private meeting between just the senator and me about five months after I arrived in Washington. I well remember McClellan asking how I liked government work. I said, rather full of myself at age twenty-three, "Not very much. No one seems to want to work very hard." Two weeks later, he called and said, "Do you still want to work?" When I said yes, he said, "Be at my office in the morning."

Thus I joined the staff of the Judiciary Subcommittee on Criminal Laws and Procedures, where hard work was indeed the fashion. After perhaps eighteen months, McClellan announced that he would not be standing for re-election. I felt I had seen enough of government-in-action, returned home, and joined Symington's law firm, not my father's.

A year later, I received another call from McClellan. He had decided to run again, after all, and could I come back to Washington, rejoin the staff, and help him with his reelection campaign? And thus I spent the next six years working directly with McClellan, serving on the staff of the Appropriations and Government Operations Committees.

There is no question that McClellan was the hardest boss I ever worked for. At our first conversation, once I was on the payroll, he had said "Tommy, I expect you to give me one hundred percent. I need for you to take this job as serious as a heart attack. But I know you are young, so don't take on anything your shoulders can't carry."

Talk about a motivator. Do you think there would ever be something I couldn't "take on"? The truth was, I was scared spitless, but I learned that working hard was the best remedy. I also learned—a lesson for the future—that expecting superior performance from the people you work with leads to superior performance. (If it doesn't, well, you may want to make some changes in staff.) From the beginning, I was challenged to take on responsibilities that I would have thought were beyond my abilities. I gained skills, and with skills came confidence, and with confidence came a smoothing of some of my rough edges.

I worked six days a week and averaged about fourteen hours a day. I reported to McClellan's number one staffer, Jim Calloway, who became my teacher and protector. We met every morning for breakfast and again at 5:00 PM when I would get my assignment to be completed for the next day's breakfast.

Whether working on a speech or laying the groundwork for legislation, McClellan was demanding, cowboy tough, resolute, honest, fair, and wise. I can cover half of those—demanding, tough, and wise—in one example. I wrote drafts for speeches and reports. When summoned for a review, I was required to bring a dictionary and a thesaurus and be prepared to defend my choice of words or phrasing. You can bet, after the first such discussion, I ran my own dictionary/thesaurus review before each meeting. This may not have materially reduced his challenges, but it certainly prepared me for the combat. Sometimes he didn't question my choices, but that I used any words at all. "Why do you have this here?" he would ask and before I could answer, he would add, "Take it out, we don't need it. The audience will appreciate some brevity." He also made me read a speech aloud, and then read it backwards, aloud. This may seem bizarre, but I learned to appreciate the cadence, the rhythm of the words, and the logic of a fluid structure.

Among other projects, I worked on the Organized Crime Control Act, the Omnibus Crime Control and Safe Streets Acts of 1968 and 1970, and the Abe Fortas Supreme Court debacle where McClellan was opposed to the 1968 nomination of Justice Fortas to be chief justice, and the staff helped assemble data to show why. McClellan and other senators staged a filibuster and President Lyndon Johnson pulled the nomination. There were various Vietnam War crises, and, of course, Watergate, where I played a modest role in the selection of Senator Sam Ervin (D-North Carolina) as chairman of the 1973 Watergate Committee hearings. (For the record, Fred D. Thompson—then, and still, my age—was minority counsel.)

But it was my work on the Appropriations Committee that set me up for the future. I had responsibility for all requests from the Department of Defense (DoD) and individual military services for research and development (R&D) appropriations, and had the appropriate security clearances and "special access." I had on-going contacts with DoD officials, the service secretaries, senior officers, top-level staff at the National Security Agency (NSA) and, especially, the Central Intelligence Agency (CIA).

I had a front row seat when issues were publicly debated and a backroom seat giving advice and counsel. In any meeting you were allowed to give your personal opinion, but only once—so you better have thought it through. Once a decision was made, you followed orders to the hilt. Perfection was a standard, not a goal. Never did you discuss with anyone outside the staff what went on in the backroom or offer what another member privately discussed when with the Old Man—a nickname never used in his presence or with outsiders. The only acceptable forms of address were "Mr. Chairman" or "Chairman McClellan." I think his first name was John. I don't think I ever heard anyone below the grade of senator call him "John."

I probably could have stayed forever (some committee staffers enjoyed lifetime tenure) but by 1975 I was ready to try something else. I arranged a lateral transfer to a less hectic government post, at the General Services Administration (GSA) center in Kansas City, and began the job search. Since I had gained a great deal of experience on the appropriations side of the senate, particularly on military programs, I assumed that something in the legal department of a defense contractor might be a good slot. I had conversations with Teledyne Ryan Aeronautical, came to an agreement, and was about to head off to San Diego until I got a call from an executive at McDonnell Douglas in St. Louis. They were interested in having someone with my background on staff. Could I come over for a meeting? Company founder, James S. McDonnell, got my name from someone and "Mr. Mac would very much like to talk with you."

Of course I went, but there was a short detour on my way to meeting Mr. Mac. Since this was basically a job interview, my visit was coordinated by the human resources department, which had me set up for a few preliminaries such as an IQ test ("Two trains leave stations fifty miles apart. . ."), a personality inventory ("Would you rather lead a symphony orchestra or build a tree house?"), and I had a meeting with the general counsel that was basically a friendly chat about "What was it like working for the Senate?"

Then I was ushered in to see Mr. Mac for another friendly chat, with no traditional interview questions. Mr. Mac wanted to know what I had been doing lately and soon enough said, straight out, that he thought I would be a good fit for the company. He asked what my government salary was. I was at the top of the scale in my job category $21,000 and he said he would match it.

So, a few weeks later (and after regrets to Teledyne Ryan) I reported aboard, expecting, of course, to be joining the legal staff. The HR director said, "Oh, no. You're going to marketing, over with the space and missile guys in the Astronautics Company." I learned I had indeed been slated for legal, until my friendly "what have you been doing lately" chat with Mr. Mac. "Lately" I had been working on some classified aerospace programs—the B-1 bomber and stealth technology. McDonnell Douglas needed someone with connections and clearances, especially the latter. Security clearances don't automatically transfer from one job to the next, but since I had been through the background checks and already had the top-level tickets, getting my clearances renewed was not a problem, nor did it take much time.

At this point, I knew something about the law, something about human nature, and a great deal about the legislative and appropriations process. I did not know anything about selling, which—contacts and clearances aside—turned out to be the most valuable attribute I brought to the job. I was not a prisoner of standard techniques or tactics; I didn't know what they were supposed to be and along the way got to invent some of my own.

As for learning about the McDonnell Douglas business in general, their history, products, competitors, triumphs, and disappointments, I had a good tutor in Mr. Mac, with whom I would have dinner every week or so. In return, I added to his store of knowledge. He was that rare executive who admitted what he didn't know and wanted to fill in the blanks. He wanted to know how "Washington" worked, who were the key insiders, how might they be influenced, and the interface of politics and the Pentagon.

I earned my pay, and justified Mr. Mac's judgment, on my first assignment. The Harpoon anti-ship missile program was in jeopardy. There was talk that the House Appropriations Committee might not fund the next stage of development. I went back to D.C., met with my former colleagues on the Senate staff, and briefed the appropriate people on the House side. I cleared up some misunderstandings, especially misguided efforts to compare the cost of the missile (then roughly half-a-million dollars) against traditional torpedoes and naval gunfire. Harpoon was designed to come at a target from a great distance, skim just above the waves to avoid enemy radar, then, at the last moment, go vertical, pop up over the target, and come down against unprotected superstructures and decks. This was one weapon doing the job of a multitude, with unsurpassed accuracy and effectiveness. It was a half-million-dollar package to take out a multimillion dollar ship long before it might come into range of guns and torpedoes and long before it could take our own ship—the launch platform—under fire. The funding was approved. Harpoon entered service two years later and remains a mainstay today.

When I was younger and really didn't know what I wanted to be when I grew up, I followed a career path that was dictated for me. I suited up every day, was conscientious and good at my job, but I was unsatisfied. When I "sold" the Harpoon, I knew that I was at the right place doing the right thing. Over time, as I grew into the job, I saw how the business of selling missiles and airplanes worked, or, more often to the point, didn't work, and began to develop the processes and procedures that are at the heart of this book.

CHAPTER 1

The Business of Aerospace

In the late 1970s, the world of American military aerospace was conflicted with a stagnant defense budget (the post-Vietnam "Peace Dividend") and the traditional sales model—using retired fighter pilots with friends still on active duty—seemed as a good as any. By the1980s, the Reagan defense build-up offered greater opportunities waiting to be explored (or exploited), but the nature of the product was radically changing. An industry created by entrepreneurs had outpaced itself and was being overtaken by technology: electronics, microcircuits, advanced weapons, and more powerful engines. With greatly increased cost, selling had to be more sophisticated. It was over this transition, from the traditional to the unknown, and when our companies were losing programs to competitors left and right, that I entered the world of aerospace marketing.

Some Basics

To sell military airplanes, you need to know a lot more about airplanes than range, ceiling, speed, weapons, maintenance schedules, and asking price. You need to know who would want to buy some, and why, and how they might pay for them. You also need to know about the probable competition and about the intricacies of industrial participation, better known as "offset," whereby you offer in some fashion to return a portion of the purchase price to the customer in the form of local jobs or investments. It also helps to know something about building airplanes in general, and military airplanes in particular. We'll start with this last item first and cover the others on a need-to-know basis throughout the book.

The rough timeline for the design, development, and manufacture of a modern military fighter begins with a bright idea and a clean sheet of paper. Plan on five years to complete System Design and Development (originally called "Full Scale Development" in my day). To begin production, allow eighteen months for receipt of long-lead materials and parts and then another two years to build and assemble an airplane. Adding it all together, it takes almost nine years, from start to finish, to get the first copy. More or less. This does not include stops and re-starts, challenges from Congress and competitors, or changes mandated in testing or because of the availability of some new must-have capability.

The timeline for the development and introduction of new materials—a new high-temperature alloy or light-weight composite, perhaps—is in the neighborhood of fifteen years, from "Ah-ha!" to approval for use in any key component of a man-rated aircraft, that is, one piloted by or carrying a human. Have patience. A lot of people with PhDs in materials engineering retire before their greatest contribution ever reaches a production airplane.

Some of the McDonnell Douglas airplanes you will be reading about in this book are:

- F-15A Eagle. First delivered in 1974 (upgraded through several versions to the F-15E Strike Eagle, 1988). The premier air-superiority fighter of its day, it kept the airspace above the battlefield clear of enemy aircraft so the attack and transport airplanes and helicopters could work unimpeded. The F-15E added ground attack capabilities.
- F-18 Hornet. A carrier-based Navy fighter with attack-aircraft capabilities that first entered service in 1983, jointly produced with Northrop. The initial Navy designation was F-18, later changed to F/A-18 (pronounced "effenay-18") to reflect the dual fighter/attack role, but the plane was usually sold internationally as F-18, often with some of the attack features removed. To keep things simple, I'll just call it F-18.

- C-17 Globemaster II Long-Range Airlifter. Operational in 1995, it was designed to take such out-size cargo as tanks, trucks, and large artillery pieces into unimproved airfields that are not accessible to larger airlifters (and far more cargo overall than could be delivered by the smaller airlifters that were designed for the unimproved strips).
- AH-64A Apache Attack Helicopter. Developed in the early 1980s by Hughes Helicopters (purchased by McDonnell Douglas in 1984) and later upgraded (from 1997) to AH-64D "Longbow" Apache with new fire control radar and the Longbow Hellfire air-to-ground missile.
- AV-8A Harrier, later upgraded to AV-8B Harrier II Plus. The V/STOL (vertical/short take off and landing) attack airplane that can take off, hover, and land like a helicopter and almost reach the speed of sound in level flight. Originally developed by a British company, it was upgraded by McDonnell Douglas and jointly produced with companies in Britain, Spain, and Italy.
- MD-11 three-engine, long-range commercial airliner in the 240-400 seat class.

How do you know when it's time to come out with a new fighter? The answer is not, "When the threat has grown and needs to be contained." During one of our early mentoring sessions, McDonnell Douglas company founder, James S. McDonnell (referred to as Mr. Mac by those around him), explained his rule of thumb. "I want to be two and a half generations ahead of the enemy. And, of course, the competition." At the same time, he added, "You don't want to bring out a new model too early. How do you know what is the right time? You can break an airplane down into five major components: the planform (hull and empennage), structures, avionics, propulsion, and weapons. I want to have a breakthrough in three of the five before developing a new airplane." Move too early, you'll lose. "I can't tell you," he said, "how many airplanes never went anywhere because they didn't offer enough advantage over existing models."

Mr. Mac also warned that there always will be one airplane flying somewhere that no one knows about, a product of a secret development project. That's the one that will bite you in the order book. Mr. Mac had another rule: McDonnell Douglas built twin-engine fighters, and much of our competition came from single-engine airplanes which, on balance, cost less to build and were priced accordingly. Customers clearly liked single-engine products, why didn't McDonnell Douglas offer one?

There's a simple explanation. Beyond the obvious (having double the power affects speed, climb rate, and, above all, the two-engine margin of safety) more power means you start with a bigger platform than you need at the time. This leaves room to accommodate whatever improved electronics, new weapons or systems that may come along in the future, without having to change the basics. Mr. Mac told his designers to create a platform that provided space to grow. He knew it would eventually be filled, and the engines would still have enough margin to handle the increased load.

The large platform made it easier to upgrade, to come out with an improved model, or to switch to improved engines as they became available every ten years or so. We took our air-supremacy fighter, the F-15, from F-15A to F-15E over a period of sixteen years. Of course, all manufacturers work to improve their products, sometimes identified with a "block" designation. You might read about Brand X Block 40/42 or 50/52, designating two late model sequential upgrades in single or two-seat (usually training) aircraft.

Whatever and whenever, you design the airplane you need for the day after tomorrow's threat, as best it can be determined. Improved radar, more agile missiles, exotica such as lasers or high frequency sound waves, who knows what dangers might be lurking in the future? The latest U.S. fighters, the F-22 and the F-35, could take on the world. However, here's a current and major challenge: most of our known or likely enemies don't bother with sophisticated systems. Few have airplanes at all. They do their fighting on the ground. To counter improvised explosive devices and suicide bombers you need more brute force than high technology. This opens the market for attack helicopters and air-to-ground attack planes to take out safe havens, meeting places, supply lines, and bomb-making factories.

To help make sense of much of what you will be reading, here's a three-minute, simplified, tutorial about how governments buy airplanes and about restraints on the sort of inducements that you might offer to sell them, especially in the international marketplace.

Let's say that the Department of Defense (DoD) or one of the services has identified a need—time for a new fighter. There will be a lot of meetings, interaction with contractors, questions asked and answered, resulting in a Request for Proposals (RFP). More questions asked, and answered, perhaps resulting in a modified RFP. All companies will enter their proposals. The government will study the proposals and often give everyone a chance to submit a "Best and Final Offer" (BAFO). This does not necessarily mean you have to drop your price (although that may help) but is realistically intended to give everyone a chance to modify details to catch up, perhaps, with last-minute indications that certain aspects are more important to the customer than others.

For a really big program, the RFP may be pointed at bringing several companies into the game, with government funding to build prototypes and hold a competition, a fly-off to determine the winner. For most programs, there's only one winning proposal: one bidder (which may be a team) gets the contract to build the product or provide the service.

Or, perhaps your company has a great idea for a new product that will revolutionize warfare (or, at the least, add capability, save money, whatever). You can make an unsolicited proposal seeking research and development (R&D) funding. Or you can provide your own R&D until the product is ready to launch, and then seek a contract to produce.

One key to winning, of course, is to have a better offer than the next guy, with some combination of better features, better benefits, and better price. At times, there may be factors that have little to do with the actual merits. For example, DoD may want to ensure that too many contracts don't go to just one bidder, and spread the work around to preserve competitiveness and the industrial base.

These comments do not necessarily apply to international competitions as each country may have its own rules or no discernable rules at all. A couple of dozen guys sitting on an oil well who want to start an air force and read about your company in the *New York Times* may just give you a call and write a check.

On the subject of "inducements"—in plain language, offering a bribe to make a sale—domestic ground rules have long been pretty well understood:

don't. However, for many years, there were no rules governing international sales, and some parts of the world were like a lawless frontier where he who pays the most, most often wins. In 1977, Congress learned that more than four hundred American companies had recently made questionable or illegal payments in excess of 300 million dollars to foreign government officials, politicians, and political parties, and stepped in to call "Halt!" with the Foreign Corrupt Practices Act (FCPA). The Act prohibits U.S. companies and their agents from giving bribes, finder's fees, and excessive consulting fees to win foreign contracts.

The amount of the inducement is immaterial; ten cents or ten thousand dollars, it's all the same. The individual making the offer may go to jail for as long as five years and be fined up to 100,000 dollars, as an individual. On top of that, the fine *may not* be paid by an employer, which could be separately fined several million dollars.

The rules are complicated. For example, under most circumstances, you can't pay a fee to a government official, and you can't pay a fee to someone connected with the government. How connected is "connected"? There was a time when some companies saw the ideal consultant as a king's brother, who was paid a princely salary to do nothing except be the king's brother. No longer. But suppose you want to hire a consultant who has a great contact with the deputy prime minister. If that means he went to school with the deputy prime minister, that's okay. If the deputy PM is his brother-in-law, it's not okay, because he is too "connected."

What about, other than "most" circumstances? It is okay to pay a fee to an official to get a permit, or have your mail delivered, or process a visa, and so forth, if these are duties he is already obligated to perform. You just want to get him to perform them a bit faster, in which case the fee is considered a "grease payment." At McDonnell Douglas, I had two lawyers always on call, assigned to parse this sort of thing, and to approve, or not, any ambiguous transactions.

FCPA was designed to restore public confidence in the integrity of business and to level the playing field, at least among American companies. If anyone ever came in to me and said, "We can't win this without breaking the rules," he was gone. As far as I was concerned, he was just making up excuses for not trying. Now, if he tried and lost because the competition broke the rules, at least he stayed ethical. As you will appreciate, some companies were unhappy with the FCPA, and at one point, started a move to have Congress amend the Act, weaken it, and allow more wiggle room. McDonnell Douglas was on the other side of that effort: We argued that everyone benefited when all played by the rules—clear, understandable, unambiguous rules. We issued a statement to that effect, testified on the Hill, and took guff from some members of Congress because their constituents wanted it changed.

During my active selling years, foreign competitors were not so constrained. What should you do when you discover that every member of the source-selection team is all-of-a-sudden driving identical brand-new Mercedes-Benz sedans? You have a quiet talk with your customer, explain the rules under which you must operate, and reinforce the superior qualities of your offering.

Things improved in 1997, when the Organization of Economic Cooperation and Development (OECD) adopted a "Convention on Combating Bribery of Foreign Public Officials in International Business Transactions." This put 33 nations on the same page, more or less. Of course, neither the FCPA nor the Convention inhibits the "customer" from suggesting creative arrangements, "you pay me this and I will do that." Forewarned is forearmed.

One Friday evening, after I'd been handling marketing for the astronautics company for about eighteen months and was having dinner with Mr. Mac, he said, "Oh, by the way, Monday don't go back to your job. . . ." My heart sank; I thought I'd been doing very well, and now? But he continued, "Come over to corporate headquarters, I've got an office for you down the hall." As soon as I got home, I called my boss, not sure what to say, but he said, "I know all about it, it's okay, come by and pick up your stuff. It's a great move up."

Indeed. I shared an office and a secretary with Mr. Mac's oldest son, Jim, who was the corporate vice president for International, and worked my way through a couple of different job titles—special assistant, director of government affairs, and director of marketing operations and strategic planning—before heading off as staff vice president for Washington operations in 1983.

From my first days with the company, I was privileged to attend a weekly routine meeting of the senior leadership 7:30 every Saturday morning, deep in a World War II bomb shelter. Present, perhaps six or seven people—the corporate CEO, the aircraft company president and executive vice president, the head of marketing, the chief financial guy and a couple of others—who would run through all of the on-going competitions, where we stood, and what we needed to do next. The meetings would start promptly and end just in time for most of the attendees to make a 10 AM tee time at the nearest golf course. While I was not yet part of any senior leadership, I was included in the meetings to offer insights into the world of D.C. politics and Congress, to take action notes and to report, week to week, progress on one item or another.

It was at, I think, my third meeting, 1975, that I learned that one of the senior executives had turned down an offer from the Air Force to become involved in the highly-classified development of stealth technology. I already knew a lot about stealth research from my work with the Senate, and I was flabbergasted, and said so. I certainly didn't get off to a good start with *that* executive.

Stealth is a magical blend of curvy planform (to reduce radar-reflecting surfaces) and special coatings (to absorb much of the radar energy that reaches the airframe) that can render an airplane almost invisible to radar, or, at least shrink the radar image of a large fighter to the size of a sparrow. Stealth could be applied to bombers, helicopters, missiles, tanks, warships, and make radar as we knew it impotent. Getting into stealth would have cost a bundle of money, and that executive didn't appreciate the value. We were too busy, anyway, he said, and had too much on our plate. Thus, for a time, the world's leading designer of fighter aircraft abandoned the field to others, notably Lockheed, which was soon awarded the contract to develop and build the F-117 stealth fighter.

At some point later, however, McDonnell Douglas became deeply involved in stealth. Not everyone in the company was tone deaf. It was too late to be in on the ground floor, but not too late to make meaningful contributions to the state of the art.

How to get into the winner's circle is, of course, our topic for today. My informal education began with *The Anatomy of a Win: Aerospace Marketing for Aerospace Management* by James M. Beveridge. I don't remember where I found the book—it may have been in a used bookstore, or a friend may have given it to me—but I certainly remember the thrill of discovery and spent many hours underlining key passages with red pencil. "Engineers make lousy marketers. . . . Engineers should engineer. Manufacturing people should manufacture. Programmers should program." "A superior proposal is necessary, but it doesn't insure a win. A poor proposal insures a loss." When giving a presentation, "Never, never" read your charts verbatim; "What is more boring than seeing a chart and then having the speaker read each line to you?" (James M. Beveridge, *The Anatomy of a Win: Aerospace Marketing for Aerospace Management.* Playa del Rey, CA: JMBeveridge and Associates, Inc. 1964.)

Here was a book written by an aerospace marketing pioneer, a man who had honed his skills while working at Hughes Aircraft Company, and I was the ideal student: the eager neophyte, willing to learn about marketing, working with customers, influencing the influencers, deciding when to bid on a contract—and when not. It's un-common sense for people too busy to think for themselves.

At some point, I dropped it on the desk of Don Malvern, F-15 program manager, and said, "Here are some of the things I would plan to do if I ever get the chance to be in charge of something." Yes, working with McClellan had smoothed some of my rough edges but had done little to stifle my arrogance. I would like to think it was the arrogance of youth, except I was now over thirty, so it was just, well, arrogance. I soon enough learned how Malvern reacted, not just to that conversation, but to Tom Gunn in general. "I don't like you," he told me. Since he was on track to becoming the president of the company, this was not a good sign. Not too long after, he was hospitalized for some reason just as a crisis was brewing with the threat of a congressional committee hearing over some F-15 issue. I fixed the problem. The next time I saw him, he said, "I've changed my mind. I'll always be your friend."

Don passed away just about the time I started writing this book. He was, indeed, my friend and still had the copy of *The Anatomy of a Win.* His widow returned the book to me and said, "Don told me that you asked him to read it because it showed how you intended to run marketing once the company got smart and put you in charge!"

Beveridge caught my interest and set the stage, but as I would soon learn, by the time I read his book (nine years after it had been published) the world of aerospace marketing had grown more complex. Yet, our company had no comprehensive new business plan or process and handled each opportunity ad hoc. We needed a common, disciplined approach in all business areas, with business plans, strategies, thorough competitive analyses, and a determination of the likely cost of entering a competition including, for example, the number of man-hours at what billable rates would be needed to develop the proposal. That could be a substantial figure, say, 20,000 dollars an hour for a large team for maybe six months. Not something you undertake on a whim or a hope.

Getting Organized

In 1986, after a three-year tour in the Washington office, I returned to St. Louis as vice president of marketing, with the opportunity to implement some of my ideas. I began by holding "Plans and Strategy Reviews," wherein the program marketing managers would try to prove that they knew what they were doing. It was rather like a confrontational congressional hearing. The senior marketing staff and I would sit behind a line of tables in the front of the room, the hot seat was at a table facing ours, and the entire department, around a hundred people, would sit in the back and absorb wisdom.

Participants learned pretty quickly what we wanted to hear, and what we did not want to hear. In the first meeting, I must have said the same thing a dozen times: "I don't give a f___ what you think, I want to know what you know!" The first victim was a young engineer, just hired into marketing. A head-and-shoulders type, first in his class at a major university, but not yet ready for prime-time. I interrupted his first briefing to ask a question, and he responded, "I think . . ." Not the right answer. I said, "For Christ's sake, three months ago you were walking around the plant with a slide rule in your pocket and wearing white socks. I don't care what you think. I want to know what you know." Sure, temporarily de-motivating, but as McClellan did with me and the Marine Corps does with new recruits (tearing them down to build them back up), motivation returns stronger and better. This particular engineer survived to become, among other postings, a vice president and program general manager. He was, of course, not my only target. Once in a while I varied the tune: "Do you mean to suggest that everything before the 'but' is bullshit?" or "Hope is not a strategy."

Ah, but what *is* a "strategy"? Here's a test: ask some of your coworkers (or, if you are a risk taker, your boss) to define the term. Watch while eyes glaze over as they go temporarily brain dead. Oh, soon enough you will get answers. Some will say it's a mission statement, some think it's a vision statement, some will describe a series of tactical moves. Most will likely miss the point: "strategy" must lead to a goal. Win the war, win the contract. In essence, you are posing the question, "What will it take to win?" and providing the answers. Good price. Support from the user community. Solid offset proposal. If you don't have a goal, why bother?

At least once a session I would make everyone get up and go over to the windows overlooking the employee parking lot. "Do you see those cars? That's what you're working for. You put food on the tables and cars in the parking lot by selling airplanes." I passed out tie-clips with the slogan, "YCSASOYA." These always elicited the obvious, "Ah, what . . . ?" Answer: "You Can't Sell Airplanes Sitting On Your Ass." I wish I could claim that as an original idea but I stole it from a marketing guy at Douglas.

At one meeting, the agenda included the Korean Fighter Program, a long-running effort to sell F-18s to Korea. The program was not going well, and the resident marketing managers knew it, and knew that the briefing would become ugly. So they persuaded an innocent subordinate to make the presentation. He started by telling us that his position reminded him of a tale from "Uncle Remus" wherein a bear is hanging upside down from a tree, tied at the feet. He's there because a crafty rabbit had convinced him that by hanging there he would be making "a dollar a minute" scaring crows away from Bre'r Fox's peanut patch. Huh? I could feel the blood rising to my face as the briefer went on to highlight a list of strategy objectives that had been adopted but not accomplished, and then laid out a series of steps he thought we should take to recover.

I went loudly crazy. The briefer got kind of crazy back—he stood his ground and defended his position. As I later learned (when we promoted him to run the Korea sales program), he thought he was going to be fired on the spot. The promotion was a good choice. After years of stumbling around, we won. In December 1989 the Koreans agreed to buy 120 F-18s. (A year later they cancelled the deal, for a variety of reasons, some of which made sense and some of which were angering. See page 117.)

McDonnell Douglas proposals had always been hip-pocket affairs, put together by teams assembled for the purpose with appropriate representatives from various disciplines: marketing, engineering, finance, product support, and so forth. They sometimes won, sometimes stumbled, and there seemed to be no continuity from one proposal to the next. Eventually, influenced by the teaching of Jim Beveridge, we realized that winging it, even for an aerospace company, didn't make sense. Since we had aeronautical engineers designing our airplanes, shouldn't we have professional marketers and writers doing the proposals that allowed those engineers to remain employed?

In 1983—still in the "hip-pocket" era—the company began working on one of the more important programs of the decade, the Navy's next attack plane, the top-secret A-12 Avenger II. Al Gingrich was given the job of proposal manager. He had been educated in aeronautical and mechanical engineering but had spent the past twenty years in marketing, where he worked on a few proposals and came to know a lot about customers and their requirements. He knew engineering speak but more to the point, he knew what it took to sell things.

When he got the assignment, the program manager called him in and said, "You are the proposal manager." Pause. "Do you know what that means?"

"Yes, I manage the proposal."

"Exactly. You will tell people what goes in, when it goes in, and how it goes in. You set the schedule. If I'm ever unhappy with what you're doing, I'll let you know. Otherwise, just keep on going." A good start for any job.

It certainly was not only the largest proposal Al had ever handled, but one of the largest, ever, for the company. It filled 110 hardbound books, ranging in size from 100 to 200 pages, all researched, written, illustrated, reviewed, printed, and bound in total secrecy. McDonnell Douglas (teamed with General Dynamics) won the contract, and the proposal was rated by the Navy customer as the best they'd ever seen. Because of program classification we couldn't much brag about this but we recognized a good thing when we saw it, and Al Gingrich was tapped to create our first-ever designated "professional" proposal team.

Al hired a couple former newspapermen—a writer and an editor—added some in-house talent, and set forth to do battle. The team practiced on several contractor R&D proposals—seeking money to fund promising approaches—and won them all. These were concentrated in the 100,000–300,000 dollar range, with only a couple going above 1 million dollars. One reason for 100 percent success was a selectivity on what to propose in the first place, which later became a key element in our formal business development process.

The team's first assignment outside of St. Louis was to assist the McDonnell Douglas Helicopter Company in Arizona—just purchased from Hughes—with a proposal for the Army's Light Helicopter Experimental (LHX), a demonstration-validation competition to develop the armed reconnaissance helicopter of the future. The team was met with a mix of suspicion and amusement. How could five outsiders from St. Louis, of all places, possibly come and write a winning proposal when they were not only not part of the program, but didn't even work for the division? Pretty well, as it turned out. They not only wrote a winning proposal and, with partner Bell Helicopter, became one of two competing teams (Boeing/Sikorsky was the other) but also earned an accolade from the Army's program office for another "best ever seen."

Soon enough, the proposal team became the New Business Center, with a larger staff of full-time writers, editors, graphics specialists, and consultants coming in and out as the work load required, available to work with any program. Each program had a dedicated "war room," where all of the materials being worked on were available for review by everyone involved. Key points were put into storyboards, showing every step to be taken and by whom, along with a rough cut on graphics, and hung on the walls around the room. (The first efforts at storyboarding were clumsy—too much information on each sheet, almost duplicating what was in the proposal. That got trimmed back to one or two pages per section, just key points and selected graphics.)

It took some time, however, to educate folks in the value of the Center. A program team would arrive because their boss told them to hustle over to the Center to be assigned a work room; they were left pretty much alone with the request for proposal (RFP). Maybe half an hour later, the leader might pop out and ask, "Do you have someone who knows how to shred, break down the requirements of, an RFP?" Someone would go in to help. A while later, another question, another bit of assistance, then more, until the teams would be fully working together. And the word would drift out to other programs, "Use those people, they know what they're doing."

The Center began holding seminars to show executives what it was all about, emphasize their role, and get buy-in. Eventually attendance was expanded beyond executive row to include people from all over the company—engineers, support people, marketing people, financial people. What began as a two-hour orientation was expanded to two full days that included mock competitions where the attendees had to develop storyboards and proposals. These were sometimes competitive—as every discipline had its own approach and its own ideas of what should be in the proposal—but by the end of the two days each had come to understand what the others wanted to do, and why.

It had been the usual practice to invite senior executives to come by to review a proposal. The usual result was that some would, but on their schedule, of course, not ours. Someone proposed that we set up a Red Team, a group of subject matter and customer experts independent of the executives, to evaluate the proposal from the customer's point of view. In practice, there might be several Red Teams, perhaps five people each, one for each major section. Were we sensitive to the customer's specific requests, and did we make the case for our superior approach? Less wear and tear on busy executives, more focused results. People quickly came to realize that the Red Team was not a make work exercise for people between more important assignments. On one competition, the Red Team came up with four warning flags and said "Fix these or you will lose." They weren't fixed, and we lost. The customer reported the reasons for the loss: the four flagged items.

One of our consultants offered a ploy to get busy executives to at least notice, and perhaps even pay attention, to what was happening in the New Business Center. We hung a series of storyboards in the hallway through which all executives walked several times a day. This often prompted some good interaction; a support expert might notice something on an engineering panel and pass to his staff, "Here's something that will impact your department, check it out."

I got better at my job with each passing effort and I was greatly aided by a brilliant process guru, Richard Hodapp, whose company The Mapping Alliance marketed (then, and still) a training program and a set of decision-making tools under the trademark name, Decision Mapping. I paid a million bucks and it was worth every penny.

After almost four years back in St. Louis and with a pretty good track record, I was sent out to Mesa, Arizona, as president of the Helicopter Company. This was not a reward, but a challenge, as the Helicopter Company was broken. But as it happened, this wasn't the only business in trouble.

The Birth of the Process

In 1991, the bottom fell out like the trap-door of a gallows, leaving McDonnell Douglas gasping for breath and barely holding on to life. In January, DoD gave the LHX program to competitor Boeing/Sikorsky and cancelled our A-12 contract. In March, Korea reversed the decision to buy F-18s and gave the contract to the General Dynamics F-16. In August, the cruelest blow of all: we lost the competition for the Advanced Tactical Fighter. We lost the future. CEO John McDonnell wanted to know why.

John brought me in from Arizona, to team with Jim Dorrenbacher (president of Astronautics) and answer the question. It didn't take long to decide that the LHX loss was, simply, that—a loss, despite our submission of the best proposal that office had ever seen. As Beveridge said, "A superior proposal is necessary, but it doesn't insure a win. A poor proposal ensures a loss." Boeing/Sikorsky had been ahead on points and was awarded the contract for what was then officially named Comanche RAH-66.

In any event, however, we didn't lose all that much. We leveraged a consolation prize with approval to remanufacture and upgrade all Apache A-64A attack helicopters into a more versatile, more lethal AH-64D "Longbow" Apache. Downstream, in 2004, the Boeing/Sikorsky Comanche program was cancelled while yet today, the McDonnell Douglas Longbow sails on. (I write this from the perspective of the old McDonnell Douglas, not of the new Boeing. I trust you will understand.)

The A-12 program was incredibly complex, and pushing the state of the art to develop an ultra-stealthy flying wing led to many issues, including large unresolved costs. And blame. The secretary of defense didn't want to deal with it anymore and cancelled for no reason that I could see, but because he could. The companies sued the government for breach of contract. This book is not the forum to rehearse the dispute, especially since today, sixteen years later, litigation continues unresolved.

As for the Korean F-18 program, we had no one to blame but ourselves. We won it fair and square; we lost it because our adversaries continued to work the customer, operating as if the competition was still open. And no one on our side seemed to have been paying any attention. A case study with full details appears on page 117.

And finally, the Air Force Advanced Tactical Fighter program (ATF). This was *the* program for the next fifty years. "Advanced" was the operative word. The program included next-generation power plants with thrust-vectoring to enhance maneuverability, the latest avionics, electronics, weapons, and especially, the latest in stealth technology. The ATF was the downstream replacement for the F-15 and a counter to advancing Soviet technology. McDonnell Douglas had been delivering F-15s, then the premier air-superiority bird, since 1974. But F-15 production was winding down, and a win was critical to the future of the company.

The RFP was issued in mid-1983, and seven companies entered the fray: Boeing, General Dynamics, Grumman, Lockheed, Northrop, Rockwell, and McDonnell Douglas. At some point, the Air Force changed the head-to-head paper competition into a "dem-val"—demonstration/validation, as with the LHX program—wherein selected bidders would win the opportunity to build two airframes, each to be equipped with an engine being offered by one of two other bidders, Pratt & Whitney and General Electric. The Air Force strongly urged that the various competitors seek teaming arrangements, which would winnow down the number of bidders and simplify the process. And, who knows? A teammate might add some useful innovations.

Lockheed, Boeing, and General Dynamics quickly made a deal. Grumman and Rockwell had dropped out somewhere along the way. McDonnell Douglas hemmed and hawed—the ATF development team had the attitude, "Don't you know? WE are the fighter guys?" and didn't see a reason to share the glory. At almost the last minute, the company joined forces with Northrop, the only un-affiliated competitor left and already our partner on the F-18 program. That had been a troubled marriage but never reached the level of divorce, and Northrop indeed had a lot to offer, with solid manufacturing capabilities, solid contacts in Washington, and—most important—a solid grasp on stealth technology that was well-incorporated in their proposed ATF scheme. Add that to our own recently-developed stealth capabilities, and we should have a winner, even against the industry stealth-leader, Lockheed.

The Lockheed, Boeing, and General Dynamics team entered the YF-22; our team offered the YF-23 based on the Northrop design scheme. We built two prototypes. We demonstrated two prototypes. We lost. In a public statement, Secretary of the Air Force Donald Rice said that neither team's prototypes were more maneuverable or stealthy, but that the Air Force preferred the Lockheed team management and production plans and had greater confidence that they could produce the aircraft at projected costs. (Some fifteen years later, the F-22 program is significantly over-budget and the original proposed buy of 796 has been dropped to 181. However, this is another argument for another forum.) John McDonnell wanted to know what went wrong, and what we could do, better, in the future. Sad to say, it wasn't too hard to answer his first question. Figuring out what to do about it was a bit—a lot—more complicated.

Why did we lose? You could start (and probably end) with an arrogant program manager who alienated not only everyone on his own team but, more to the point, on the customer's selection team as well. He'd been a general in the Air Force and therefore knew what the Air Force needed, even if the Air Force didn't agree with him. He didn't like marketing people and did not bring in any until we forced his hand well into the program. Some of our design engineers couldn't settle on a scheme and wanted to offer two basic options and let the Air Force make a choice. No one was having a dialogue with the Air Force. No one was riding herd on the design engineers.

One of our USAF (retired) consultants had warned me just before the teaming exercise when there were still five active competitors, "I've been nosing around the building [Pentagon]. Unless you get him off the program, you're going to lose. If the competition were to be decided today, you'd be number four out of five." The company took heed, shifted the program manager to another job, and, eventually, showed him the exit. A bit late in the game, but it wasn't his fault. He was who he was. It was our fault, because we didn't know who he was.

Secretary Rice's politically-correct explanation notwithstanding—that the YF-22 and YF-23 were both good airplanes and that management and cost were basic discriminators—there were some technical differences. By all inside-the-beltway accounts, our YF-23 had the edge in stealth, but the YF-22 with the Pratt & Whitney thrust-vectoring engines outperformed our offering, with either engine, at slow speeds and low altitudes. The evaluators thought we might have trouble handling the integration of ATF thrust vectoring. Apparently word had not reached them that we had been experimenting with thrust vectoring for some time using an F-15 equipped with the Pratt & Whitney engine.

The YF-23 was a damn good airplane and, one off-putting personality aside, we might have come out on top but as Secretary Rice said, we lost on management, production, and cost—the heart of any written proposal, and preparing the proposal was a responsibility taken on by McDonnell Douglas. Before you jump in and say, "Wait a minute! McDonnell Douglas was just then writing best-in-class proposals." Yes. The key phrase is, "just then." The ATF and the A-12 proposals were being developed by two different teams at the same time and none of the proposal-writing improvements had yet been initiated. The A-12 proposal was a winner, but the proposal for the ATF was not. Clearly, here was our initial point of focus.

Dorrenbacher and I brought in management consultants A. T. Kearney and assembled a detailed report on the loss, highlighting the factors noted earlier, along with others of probably lesser import. Probably. But if you add one factor to the next, and the next, and so on, at what point does it reach the tipping point where the proposal goes from "in the running" to "out of gas"? We found that, in spite of the more-organized approach McDonnell Douglas had begun applying to competitions, the company still relied on tribal knowledge to train marketing managers and there still was no consistent, company-wide practice or policy. The next step was clear: create one.

A. T. Kearney helped us develop an interview questionnaire for key, mostly retired, senior business executives and not just people from McDonnell Douglas. "What," we asked, in essence, "are the critical steps to winning?" Ask the right people with a lot of solid experience and you will get some good advice, indeed.

Among other things, they suggested we start by assessing the present and future: Where is the market? Where are the emerging markets? Who are the competitors? Where might your products fit? What are the barriers to entry? How much revenue might you generate? Are you doing everything you could or should be doing to sustain or grow the business base? Are you investing your money in the right research and development?

For any given competition, start by listening to the customer. Analyze previous competitions held by the customer: delineate the discriminators, the technical issues, pricing. Assess your current competitor's strengths and weaknesses. Assess your own strengths and weaknesses, and develop remedies. Have a defined strategy. Assemble sufficient resources (personnel, facilities, money) and make sure that everyone on the team has a specific assignment and that everyone knows what everyone else is supposed to do. Read the RFP *carefully*: competitions have been lost because someone misunderstood, or chose to ignore, a clause in the document. (If the RFP says high risk with low cost is okay, high cost with low risk is okay, but high risk with high cost is a loser, don't submit a bid with high risk and high cost. Seem too obvious? I know of one competition with that exact scenario.) Finally, have a solid proposal.

Bring in Win Strategy Steering Committees (WSSC, pronounced "wissik") to provide an independent in-depth look at the mechanics of the campaign before any shortcomings might be exposed to the customer. Members can come from throughout the corporation, key suppliers, or be consultants, but all must be experienced and know the product and the customer. For international competitions, you must have some country specialists. No member of your marketing team should be a voting member of the WSSC; that would defeat the whole purpose. The chairman must be of sufficient stature to be able to report directly to senior management. The more senior the better.

I sat down and wrote the first draft of a "New Business Activity," incorporating all of the above and with finishing touches added by A. T. Kearney. We began training not just program managers but all senior officers.

After I had worked through the St. Louis crowd (to a good reception), I took my two-day dog and pony show out to brief the commercial aircraft marketers at Douglas. My first charts were met by blank, uncomprehending stares; some of the following charts provoked irritation tending toward anger. The marketers didn't care about (or understand) Pareto charts, Gantt charts, probability of win statistics, inventory control, and cost control. They just wanted to sell airplanes to traditional customers in a traditional manner. As presented in the briefing, the New Business Activity focused on the government-military market, where detail was vital and cost discipline critical; where head-to-head competitions turned on arcane minutiae; where the customer was not just an airline operating somewhere in the world, but the U.S. military operating everywhere in the world under the constant "supervision" of DoD and the Congress, with government inspectors camped out at the factory and the Government Accountability Office (GAO) running inquiries.

In those days, commercial sales were more personal, much less technical, less structured than today—less demanding. There was no "standard" model of a passenger airplane. Each airline added its own specifications from the color of the paint job to the color of the carpets. It seemed as if no two airliners coming down the line ever had the same interior configuration. Yes, there was competition, but the key elements were price, capacity, efficiency, delivery.

It didn't take me long to realize that we weren't on the same wavelength. I cut the first day's meeting short, went back to my hotel, and converted the key charts into questions relevant to the commercial business. "These are the things you need to know about your customer. How is he organized? Is he moving into new markets? What are the age-factors of the fleet? How many aircraft of what type are nearing retirement? Who is the main decision maker? Why did he buy the last batch of airliners from a competitor (price, durability, familiarity, the salesman was the chairman's nephew)?" You get the idea; the Douglas folk certainly did, as if a light went on in an "ah-ha!" moment. I wasn't trying to force them into a marketing straightjacket, I was showing them how an orderly, consistent, disciplined process would help them plan, strategize, and make sales. To keep the light on, we sent a full-time marketing coordinator to work with them, especially to build a hybrid New Business Activity that was a better fit for the commercial business, while at the same time recognizing that Douglas Aircraft was just then moving into the government-military arena with the C-17 airlifter. The C-17 would offer marketing challenges a-plenty, indeed. For more on this, see the Case Study on page 108.

A brief observation on human nature. As the New Business Activity began to take hold throughout the corporation, other people began to add value. They added so much value that it became hard to follow, top-heavy, and was collapsing under its own weight.

Time to re-tool. We slimmed it down to a basic 12 step process that coaxed you, asked you, and finally forced you to confront every key aspect of a sale, with a simple underlying theme: don't try to sell them what you have, sell them what they need.

1. WHAT is the customer's problem? Why does it need to be solved? (The air force is geriatric, a next-door neighbor is rattling sabers.)
2. HOW will the customer address the problem? What actions must the customer take? What resources must be invested in the solution, and in what time frame? (Survey the aerospace marketplace, issue an RFP, and put a few billion dollars in the defense budget.) Is this effort likely

to be related to an effort/decision made somewhere else (e.g., legislative curbs or guidelines, actions of the political opposition or direct aggression from the country next door)?
3. WHO is the person with the ultimate authority to make the decision? That is, the person whose decision cannot be overruled by anyone.
4. IDENTIFY the internal influencers who will feed data to the decision maker. Who are the people who will implement the decision? Who will be the actual users of the product or service? Who are the outside influencers—politicians, pundits, the prime minister's mother-in-law? This helps define the universe you must address and convince, in one way or another.
5. WHAT are the customer's Most Important Requirements (MIRS)—needs, wants, requirements, and expectations? You can't pull this out of thin air. You need some intelligence, have to read the newspapers (especially in the target country, and you may need to hire a good translator), talk with folks—the players themselves, your consultants, counterparts in the U.S. military, staff at the local U.S. Embassy. As a simple illustration, let's assume that the customer wants to be recognized as a world leader, wants access to technology, requires industrial participation to help offset local unemployment, and needs help in dumping the obsolete equipment that will be left when the new comes in.
6. PRIORITIZE the factors derived in step 5; give each a numeric value, as each might be perceived by the key players or influencers:
    - The prime minister wants to play world leader and ensure his place in history (give it a 100), thinks technology is useful (a 50), is concerned with unemployment (he is soon up for reelection, call it 80), and doesn't worry much about the obsolete equipment (score 10).
    - The minister of labor is all for industrial participation (an 80) and cares little about technology (20), world-leadership (20), and dumping airplanes (10).
    - The Air Force chief of staff covets technology (90) and wants to be in the world-leader column (90) and wants help with the obsolete equipment (60); unemployment? (give it 20).

Add the scores, get a ranking. In this example, with four categories (there should be as many as you can identify) and only three key players (there would be dozens), world leadership comes out on top, 210; participation/employment, 180; technology, 160; finding a new home for old airplanes, 80.

You get the idea. Prioritizing helps you frame the arguments, shape the messages, and know upon whom to put the focus about which.

7. What are the ISSUES? Can you meet all of the customer's wants, needs, requirements, and expectations or only some of them? What factors might intrude (famine, pestilence, war, mergers)? In the past, has your company had good relationships with the customer? Are there budget/funding situations? How do you think your offer will stack up against the competitors?
8. PONDER on how to resolve/overcome/mitigate the issues.
9. DRAFT THE OFFER. What will the customer receive if you are the winner? Cover all bases including but not limited to technical, management, support, pricing, delivery, warrantees, and guarantees. Address resolution of the issues and MIRS (needs, wants, requirements, expectations) as appropriate.
10. ROLE PLAYING. What are the strengths and weaknesses of your competitors as they would appear to the customer, not your in-house analysts? How do you think your competitor's offer will meet the customer's technical requirements and MIRS? Develop a Strengths, Weaknesses, Opportunities and Threats (SWOT) analysis of your position.
    - Strength: what you do well that the customer likes (e.g., demonstrate interest and concern for the customer).
    - Weakness: what the customer has asked for that you can't deliver (e.g., shortfall on a specification in the RFP).
    - Opportunities: beyond this contract, the likely downstream benefits of a win (e.g., enhanced potential for international sales).
    - Threats: things that might get in the way of a win (e.g., an election scheduled between proposal and selection).
11. WRITE UP YOUR STRATEGY FOR WINNING. Review all of the above, factor in your own requirements (such as cost, required profit

margin, production and delivery schedules). Take a hard look: what will it take for you to win? The customer wants to enjoy international respect but on the cheap, be a technology maven, conquer unemployment, and get rid of old aircraft. Lay it out with tangible specifics. For example, you will provide 100 export model state-of-the-art fighters with technology upgrades as they become available, guarantee a fixed-cost for a five-year buy, include pilot and maintenance crew training at no additional cost, set up an in-country assembly line, and buy the obsolete planes (which you will sell to someone in South America).

12. TELL THE CUSTOMER. Logically, at step 12, you tell the customer about your offer. However, there is more to Step 12 than writing and delivering a proposal and, in truth, under the rubric of "tell the customer," many elements of Step 12 would come well before even the RFP is issued. Someone—several someones—should have been talking with the customer's someones and influencers, selling your company, your skill, your history, your record of great achievement. If there are any domestic U.S. political issues, you should have been working with selected political folk. Are there political issues in the customer's homeland, perhaps an opposition party striving for power? You might want to tell them about your company, your skill, your history, and so on, and how your airplanes solve problems. You should have been working the media, especially the aerospace trade press. You should have been demonstrating your products at various international air shows and expos. And you should have invited the prime decision maker and his closest advisers for a plant visit and tour and a ride in the two-seat training version. Have you been "telling the customer" all along the way? Are you satisfied that your company is well-positioned to get a positive hearing? Time to write up, review, fix, and submit the proposal.

Throughout, objectivity is paramount. Yes, you want to win but if you shade the facts or give a partisan caste to your argument, you're on a path to failure. It may be hard to maintain objectivity when the CEO has already told you what he wants to see, but you must. A cautionary tale: on one international competition the company was so confident that it started

building airplanes before a deal was done—only to disc[over had]
been courting the wrong government official, and lost. Y[et fol]-
lowing a disciplined process, letter and spirit, but were fe[eding bad]
information into the model, and refused advice from [those who]
understood the local politics.

A recipe is only as good as the cook and a process is only as good as the processors.

CHAPTER 2

People

By now, only a few dozen pages into my book, your head may be spinning, overwhelmed by so much advice and too many cautions. I trust it will all make sense as we run through cogent examples and case studies and more, starting in this section with how to assemble your team, work with consultants and agents, deal with cultural issues around the world, and deal with Congress.

The Team

Why would anyone want a career in international business development? Let me count the ways. . . . Every day is different. You will keep your mind sharp and your heart tender. You will bum with other cool, smart, sassy, clever, and caring people who become lifelong friends and associates. You will truly understand the meaning of "fellowship" as you help—and are helped by—your teammates personally, professionally, and politically.

You will learn the business from the inside-out, leaving you well-positioned for future advancements. At the same time, you may enjoy a temporary respite from daily corporate bullshit as you visit and live in some of the places you only studied in eighth-grade geography class. Working in international business development also presents a great opportunity to expand your vocabulary and try all kinds of different foods. On food, I offer one basic rule of thumb: if the natives eat it, so should you. Call it bonding with your customer. Pickled jellyfish isn't all bad. It's rather like eating coleslaw. However, sheep eyeballs take getting used to.

You will also learn to respect and deal with fear—fear of losing a deal, fear of strangers in dark alleys, fear of missing your airline connection. You will work with serfs and kings, often on the same day, and sit at the main table next to powerful leaders and when the day is done (when the mission ends) you will have wonderful stories to share with your mate and kids. Perhaps even pen a memoir.

What sort of people should be in business development? The flip answer is people who want to be in business development because of or in spite of everything I just said. I wish there was some sort of a litmus test for great salesmen, like one developed for fighter pilots by a U.S. Navy flight surgeon. He discovered two almost universal traits among the best of them: they ran upstairs two steps at a time, and at some time in their lives had owned a motorcycle. In a similar vein, football coaches look for a running back who can see "the game in slow motion" and find a path during the violence and havoc of the scrimmage—but these searches involve actual on-the-field observation.

I do know what I looked for in a candidate: people with imagination; people who demonstrated independence, maturity, judgment, fortitude, and personal integrity; people with curiosity and objectivity; people with the sort of self-confidence needed when meeting with a skeptical head of state; even people with a chip on their shoulder, who wanted to win, to prove something to themselves or their parents or the girl back in high school who refused an invitation to go bowling. I looked for people who were willing to spend a great deal of time in the field selling billions of dollars worth of airplanes against competition from not only other American aerospace companies but also offerings from Britain, France, Sweden, Russia, or various consortia. Above all, I looked for people who shared my passion and my obsession for understanding human nature. This had nothing to do with products or technology, but instead, with how decisions get made because, after all, what is selling but getting someone to agree to buy your product? This is gritty work, not glad handing and salesman's spin.

Ours were not entry-level opportunities. Personality aside and as a general rule, we hired people who already were experts in some aspect of the business, who knew the customer, or the products, or could tell you which technology to pick, or explain the current politics and the geopolitical history of those parts of the world in which we did business. Some were good at getting new business but terrible at keeping it sold. They were motivated by the competition, not maintenance. We needed both. Some were retired military, some worked for the National Security Council or other government agencies, and one taught the Yemenis how to fly fighters. We looked for the "well-rounded" candidate, with a background in sports or community service: one was very active in the Herbert Hoover Boys Club (composed entirely of disadvantaged black children) and another was the president of the city's Foreign Exchange Student Program. Many had graduate degrees in business or international studies, and some had studied abroad. Skill in a foreign language was a plus, but skill in oral and written English was mandatory.

Intelligence? I didn't think much of the fifty-question timed tests. I preferred to study the college transcripts candidates were required to file with the employment application, which provided a broader but not always helpful look. I hired one terrific student who turned out to be D-minus in social skills; on the other hand, I hired one failing student who dropped out of college before he would be kicked out, but later got his act together and now—true story—is president of a division.

You need people who will accept responsibility and, from my personal perspective, stand-up guys who were willing to bring me bad news. You don't find this out, of course, until they prove otherwise, like the two guys who were running the failing Korea program and sent a subordinate to take my abuse and were soon off the program.

We studied submitted resumes with caution; it is difficult to verify much of the information. While the Privacy Act permits a former employer to say anything truthful about performance, to play it safe (in other words, to avoid being sued) many if not most employers will only confirm limited material, such as dates of employment and final salary. Helpful, of course, but it doesn't tell you if the employee was fired for lying, padding expense accounts, selling company secrets, or just plain incompetence.

So, the job interview is key but also must be approached with caution. Most candidates anticipate the questions and memorize the answers, assisted by scads of professional guides. A Google search for the phrase "job interview" delivers more than a million sites and Amazon.com lists more than 14,000 relevant titles, such as *301 Smart Answers to Tough Interview Questions* or *Best Answers to the 201 Most Frequently Asked Interview Questions* or even *101 Great Answers to the Toughest Interview Questions*, not to mention, *201 Best Questions to Ask on Your Interview*. (Some publishing genius must have discovered that odd integers work best in titles like these.) If you want to ask, "What was your reason for leaving that job?" You most likely will get an answer suggested in one of the "01" books.

I asked the questions, but held the answers at arms length. Often, I was not interested so much in what they said, but how they said it. Body language. Did they lean forward slightly? Nod? Did they use their hands to punctuate a point or two? Did they listen when I spoke, or did they look like they were rehearsing their next contribution? Were they uncomfortable? Nervous? Did they shift around in the chair a lot? Keep in mind, "nervous" is not unusual behavior in an important job interview, but for the jobs we had to offer, too much "nervous" could be a liability. I looked for quiet confidence. All very subjective, I admit, and to take interviewer bias out of the task, candidates were interviewed by no fewer than three members of the team they would be joining.

Don't get me wrong: A candidate who didn't prepare for an interview by whatever legal methods would be a fool. Rote answers to anticipated questions are a given; it's the answers to the follow-up questions that count. "What's your favorite book?" might be followed by "What part/character/lesson was most meaningful?" Once, during an interview with a West Point grad from South Carolina, I asked, "Who has been the most influential American?" I expected to hear the name of some president or general (or someone who was both) but got Martin Luther King Jr., and my "Why?" led to a thoughtful one-hour discussion. The candidate knew what he was talking about, was hired, and became a successful member of the team. If you want to escape the rhetorical clutches of canned answers and see if the candidate can think on his or her feet, throw in a few hypotheticals. For example: (A) You are taking your boss into the Pentagon to meet the Air Force chief of staff. The appointment is at 8:00 and you are outside his office and it is 8:05 and your boss is not there. What do you do? (B) You are on the F-18 fighter program, and your assignment is to develop new markets. You have heard that Finland will soon be ordering one hundred new fighters to replace an aging fleet, but you know that Finland has always split their military aircraft buys between the Soviet Union and European countries. What will be your strategy to get the Finns to consider the F-18?

Watch out for a candidate who displays an overly-big ego, and who uses the vertical pronoun in every other sentence. The sort of person who, once he knows a little something about a subject thinks he knows everything about it. This may be a bit of overly-pop psychology, but in my experience, big egos often camouflage low self-esteem—not a good candidate to handle a gritty competition 5,000 miles from home.

Most job candidates came in from recruiters, or on the recommendation of mutual friends, but I also scouted the home team. More than once, when I was running the D.C. office and visiting a business unit in the field, I spotted a young person, clearly bright and with the right attitude, and asked, "Would you like to come work with me?"

A typical response would be, "Thanks, but I've got a family, I'm an engineer, I want to be working on airplanes."

So I would ask, "What is your dream?"

A big, confident smile: "I want to be president of the company."

"Okay, but first?"

"I want to be on the next new fighter program."

"Stay where you are and I guarantee you'll never have a shot at president, but come with me, there's a real chance you might get there."

"By going to D.C.?"

"Yeah, or our office at Wright-Patterson in Dayton. You can skyrocket your career. I'll sign a piece of paper with you, and if you do this and do it well, you'll be able to go back to the business unit you're with now, I'll guarantee your slot. And I'm going to get agreement from your boss, and your boss's boss, and have the head of HR put a memo in your jacket. Spend three or four years in D.C., and with good evals you'll get the raises and you'll go back just where you would have been, but with this terrific extra experience."

I have to acknowledge, none of my engineering recruits ever made president of the company but a bunch made vice president, especially of aircraft programs such as the Global Strike Division, F-15 and F-18. One became VP of both the F-15 and F-18 programs. And many VPs of marketing started their careers with me in D.C.

The team will need a leader. In my judgment (and from my experience) there are hazards in putting non-salesmen in charge of sales, or a finance manager or engineer in charge of marketing. Of course, there may be exceptions to the rule—I may have read about one in the *Wall Street Journal*—but I don't think I've met any of them.

The non-salesman has missed basic training—never gone door-to-door, never been on the road living from one motel to the next, never had to spend six months in Kuala Lumpur trying to convince a man with no military training and limited skills in English that he's about to buy the wrong airplanes, and succeeded.

Finance guys look at the financials: they think, if you give it the right price, you can sell anything. Sure. But suppose the competition is offering the same "right" price, or better, what's the next move? Cut the price?

Engineers strive for perfection—it's 100 percent or nothing, the bridge will fall, the ship will roll over, or the airplane won't get off the ground. Engineers have trouble dealing with partials, in doing anything before everything is perfect. Well, sales is a lot like baseball. You can't hit all of the pitches and you won't win all of the sales. But you don't wait for the right combination of a full moon, a favoring wind, and an exhausted pitcher. You step up to the plate, take your turn, and do your best.

Before you go into protect-the-engineers mode, let me elaborate. I did indeed have a lot of engineers on my teams, but only half of them succeeded in the competitive marketing environment. To help them gain the top half, I had an aggressive de-programming program. I handed each a Rubik's Cube and said, "Go at it. Solving this is like working through a sales campaign."*

---

* Hungarian architect Ernő Rubik had invented a fiendishly difficult puzzle. Each side of a plastic cube is a different color—red, yellow, green, so on—but each side is made up of nine squares mounted on a double-jointed axis. The game: twist and turn all of the squares until the color pattern is broken up. Then try to put it back together. It can be done, but not easily, because every move affects all other elements. In the first two years of international release, 1980–1982, more than 100 million cubes were sold to a willingly masochistic public.

It was a salutary exercise, but I admit that job assignments rested more on division of labor than the ability to solve quirky problems.

Education, training, speciality, and experience aside, what should you look for in a leader? The ancient Chinese general Sun Tzu wrote, in *Art of War*, that a commander must display wisdom, credibility, tranquility, benevolence. Former General Electric chairman, Jack Welch, offered his "4E" formula: a leader has Energy, knows how to inspire and Energize others, has a competitive spirit with Edge, and knows how to Execute to produce measurable results. Niccolo Machiavelli, in his little-known 1521 work on *The Art of War* suggests that a true leader selects himself: "In the armies, and among every ten men, there must be one of more life, of more heart, or at least of more authority, who with his spirit, with his words, and with his example keeps the others firm and disposed to fight." That's the one I want on my team.

I would also add, the leader must have confidence in himself and in his subordinates. How do you recognize "confidence"? Here are a few modest suggestions. The administrative assistant missed a comma, so the boss sends the letter back for retyping instead of just inserting the comma with the same pen used to sign the letter. (I am not making this up, I could mention a former secretary of a government department but don't want to poison the well just on the off-chance that he might be selected to review my book.) That's a pretty good sign of an insecure boss. The typical argument: "everything we send out should reflect the high quality of our products and services." True. In correspondence, letters should be complete, with appropriate references, attachments, and so on and written with a tone appropriate to the purpose of the missive. They should also be reasonably grammatical. But minor pen-and-ink changes send two equally-important signals: the person signing the letter actually read it and probably runs a cost-efficient (and effective) organization.

Another issue: my way, your way, different styles, similar result. This is a common problem with an executive who has been promoted and now manages a junior employee in his old job, and expects the job to be done as he always had done it. If gently challenged, he might say, "My way was clearly successful, as validated by my promotion." Well, people may be promoted for a lot of different reasons—including getting them out of the way—but validation should rest on the current tangibles, not past performance. And by tangibles, I mean making the sales.

Another example is the manager who thinks that process requires a hierarchy of committees and sub-committees and sub-sub-committees to bring forth truth, beauty, and the keys to the kingdom. Those committees will most likely be staffed by, well, engineers and finance managers, people not trained in sales-think, who get their marketing knowledge from newspapers and their product knowledge from trade journals. The probable outcome is a pooling of ignorance, with each group reinforcing the dumb ideas of the last. If I seem too parochial, too much in favor of marketing over all other disciplines, you bet I am, especially when it comes to marketing. As Jim Beveridge wrote, "Engineers make lousy marketers. . . . Engineers should engineer. Manufacturing people should manufacture. Programmers should program." Did I include some engineers and finance people on my Team? I certainly did. They were vital additions to handle engineering and financial matters. And I do admit that when resources were stretched thin, I turned some small competitions over to the engineers. "Use this as a template," I would say, handing them a copy of the process. "And let me review as you go along." It all worked out in the end.

Once you have assembled the Team, you need to give them something to do. Psychologists write about the "identity crisis" that seems to affect almost everyone. Who am I? How did I get here? How do I get out of here? Management gurus write of "counterfeit executives," people who know that they are not really qualified for the jobs they hold and live in minor terror that they will soon be unmasked.

Workers need to know who they are and where they fit into the grand scheme of things. What is the mission? What are the steps to get there? Does everyone know who among them is responsible and accountable for sales, marketing, overseeing the process, firm backlog, public relations, schedule, and so on? And do they really know what the boss expects, or is it an "I think . . ." guessing game?

It's not hard to uncover identity dysfunction. Ask members of the Team exactly who does what. For example, "George Jones, what is his job, exactly?" or, "Who's in charge of customer contact?" If you get some blank stares, you have a problem and you can't fix it just by having meetings. You need a plot, a chart, and a team family tree. It seems so easy, and yet, whenever I have gone on a "find the lessons learned" consulting job, this was the first place I looked for evidence of failure, and with almost 100 percent certainty, found it.

I have heard many arguments about the best way to organize a multiproduct company, such as, by program (F-15, F-18), by customer (DoD, Finland), by service (Air Force, Navy), by capability (fighters, transports), by world geography (Asia, Europe), or plant layout (important programs across the street from headquarters, minor programs across town). A good case can be made for any of these but to tell the truth, I don't think the marketing team should care. We're not interested in power grabs or protecting turf, although a convenient office space is always nice, like, across the hall from the CEO. Let the company do what the company will; it will change every ten years as programs wax or wane and priorities shift. Stay out of that fight and save energy for the real battles.

I believe in a core marketing function, headed by a business development professional, able to focus on the program of the day, but with outriders parked within each program or division or whatever. When I came in as head of marketing, I found the team clustered in cubicles outside my office. I started rotating them out into the projects, where they would learn more than what time I broke for lunch. I brought them in for meetings, or poker sessions after work (not on company property, of course, but an often overlooked tool for getting to know your people).

Public relations became an increasingly important part of our business. When I joined the company, we always had some PR people tucked away in a corner of the marketing department and well out of the command loop. Whenever I came back from an executive-level meeting, my PR guy would ask, "What went on? Any issues, any problems?" He would take my answers and go back to his office to prepare whatever he needed to prepare—question and answer talking points, a draft press release, a position paper, whatever it was that I could take to the boss for comments.

At some point, a light went on: why was I a middleman? I discovered that in most military outfits, the public affairs officer (PAO) reported directly to the commander, not to the executive or chief staff officer, and not to the personnel officer. The PAO was plugged into every important meeting and always on top of the issues of today or the plans for tomorrow, ready at hand to evaluate the potential impact of anything. We changed things. Thenceforth, PR reported to the president where his advice would be direct, not filtered, and where the PR person, in a meeting or in private (as good discretion might dictate) could tell the boss who is planning to announce a major layoff on Christmas Eve, "Ah, I think you're about to make a mistake and let me tell you why." (I'm exaggerating, but not much.)

PR became a major focus of the D.C. office, which to that point didn't even have a PR component. I staffed those slots with retired military PAOs, senior officers who understood military style, equipment and operations, knew their way around the Pentagon, and had spent a career learning the trade. Once they joined our team there was, however, a learning curve: military PAOs who grew up with the mantra, "the public has a right to know," had to learn that, in the civilian world, you don't have to rush out to address every issue and that anything you say can impact the stock price—up or down. They need to study the rules and guidance of the Securities and Exchange Commission, be friendly with the corporate counsel, and enter the field of play—with care.

Of course, "answering questions" is a small part of the job, but thirty years ago the full value of a dedicated communications staff was just coming into focus. It didn't take me long to realize that a PR professional brings a lot more to the table: skill in writing, with the ability to translate arcane jargon into standard English; a sensitivity to cultural traps; a ready-made network of friends and colleagues, military, civilian, and former military turned civilian; trained (and able to train others) in how to handle an interview, especially on television.

For example, here's a PAO mini-course. If you are scheduled for a television interview, be sure to arrive early. Try to ascertain if the interviewer knows much of anything about the topic and, if not, suggest a few questions that might draw you out while on camera "to help the viewers understand the issues." Have a friendly expression. If seated, lean slightly toward the interviewer; never sit at the back of the chair. If seated in a swivel chair (provided as an underhanded tactic or boneheaded mistake by a producer, it doesn't matter which) don't swivel. Don't fiddle with your glasses, a pen, or play with a rubber band. Don't try to deliver a lecture; keep your answers short. Anticipate questions and think through your answers in advance. For really critical topics, before you even step foot in the studio, have a "murder board" session, where you sit in a room full of clever colleagues who toss you every dirty question they can think of. Practice makes perfect.

And never, never, never, say, "No comment." I know, I said you don't have to address every issue. But you have to more or less answer every question. Less is sometimes more useful. "I can't comment on pending litigation." "I haven't seen that report." "I'm afraid that such personnel matters come under the privacy act; I'd be in violation if I said anything." "I'm not sure; let me get back to you on that."

By the way, you can learn all of this by attending a media training session run by professional media trainers. The cost may be 3,000 dollars and though I just saved you the money, you still have to practice.

Hire good people, give them challenging jobs, and compensate them royally. Base salary and benefits should only be part of the equation. At higher grade levels—director, vice president—an "incentive compensation" program (IC) is included as part of the pay package, but anyone, of whatever grade level, should be eligible for a bonus of some sort. This is especially true in sales. The McDonnell family believed in rewarding outstanding performance and providing incentives to help grow the business, especially to senior managers in the business units. In turn, those senior managers were empowered to nominate outstanding engineers and marketers for inclusion in the IC program, which set clearly defined goals (typically, some combination of: cash flow; revenue; return on investment, equity, and capital; and additions to firm backlog—airplanes on order for delivery a couple of years out). If you were in the program, you were given specific annual goals, some personal, some based on overall corporate performance—linked to whatever the CEO or the board of directors had promised (and delivered) to shareholders. Your IC might have accounted for 40 percent of your annual bonus with the rest based on evaluations and intangibles. And anyone who made an important contribution to a big sale might find 10,000 or 20,000 dollars in their Christmas stocking.

We had a locked room where the executives did the assignment and rotation planning for the professional staff. The room was kept locked, because the photos and resumes of all the candidates were posted on the wall, along with a nominal ranking, where each stood in the pack. We'd try to find challenging assignments for guys near the bottom of the list, to give them a chance to develop, improve. They might benefit from a fresh start with a different team or perhaps a position better suited to their temperament. A decent person but an introverted bookkeeper type struggling to be an affable salesman could be shifted to the finance team counting beans. Guys near the top of the list were being considered for special assignments or promotion.

Here's a small story, a bit off-topic but important. At one session, I wanted to nominate a brilliant African American engineer for the Congressional Fellows program in DC, specifically to work with the Black Caucus, help them better understand some national defense issues. The senior executive in the room said, "Oh, no, I don't think so. Next?" Later, meeting over and we were leaving for the day, the boss leaned over to me and said, "Why were you pushing that guy? You ever take a close look? He looks like he just came down from the trees!" Talk about lighting the fuse, I grabbed him and would have throttled him except a guy stepped in to break us up. Could I have been charged with assault? Sure. If the executive had been dumb enough to lodge charges. I'd have demanded a jury trial. I had the witnesses. Did I display the best judgment? Of course not.

That particular senior executive was later relieved of his duties, fired, for a lot of reasons. I had the pleasure of delivering the message. He had been on his way to the Middle East for a meeting, I intercepted his travel at the airport in Shannon, Ireland, gave him the good news, swapped airplanes (my pedestrian model for his executive special) so he could return to St. Louis to be dethroned and I could go on to the meeting in his stead and in high style.

McDonnell Douglas did not have employment contracts, so every worker came under the rubric "hire at will, fire at will." Thus, we were not required to provide compensation or relief to a released employee. However, McDonnell Douglas was a big company, which provided a broad field in which to find a compatible position for someone who did not work out in

Business Development. In truth, I had fairly high turnover in my department. One of my associates complained that "we rode them hard and put them up wet." Some sort of horse metaphor, I presume. As I noted earlier, we had high standards and our people also had to be loyal, driven by a sense of urgency, and be highly competitive. These are things that are hard to discern in the job interview but show up pretty quickly in the field. If they proved lacking in any of these areas, the employee would be released for a transfer, or to try their chances with another employer.

For the most part, the personnel department accepted my challenge to find another home for the failed hire. Failing that, they might offer "out placement" assistance, which included specific personal advice and career counsel coupled with the services of an outside recruiter with the guarantee that the employee would receive at least one legitimate job offer.

I wanted to be fair but I did not confuse liking someone with making an accurate evaluation of business acumen. Being fair also meant being honest and direct. My predecessor had a problem with handling "confrontation" (and what is a firing but confrontation?) and used the old "going to transfer you to Siberia" routine, expecting that the employee would refuse the transfer and quit. If the employee accepted the transfer, at least they were out of his hair. The tactic was unfair, of course, to the employee, and grossly unfair to the good, hard-working, loyal people who happened to be working in Siberia through no fault of their own.

For the record, there were two exceptions to all of the above. First, the company must conform to specific laws or regulations that apply to termination of the handicapped or aging (think, above 50) employee. And, second, there are occasions when you want to provide compensation—a severance package—not so much as a reward for past service but as inoculation against future nastiness. The package typically includes a letter that says, accept the compensation and you willingly relinquish any rights you may have for legal action in the future. And, oh, by the way, you agree to a "non-compete" clause, whereby for some period of time, measured in years, you will not take employment in any form with any company that competes with your about-to-be former employer.

When John McDonnell sent me out to run the faltering Helicopter Company, he gave me a simple charge: take over and turn it around. I determined that most of the in-place management team had to go. They were too much "in-place," too locked in to the way they did business and too comfortable working with each other. Some could be moved to other divisions, while others were eligible for retirement. I don't recall each situation, but I certainly remember one, where there was an older executive who was well-regarded but clearly upset that such a young whippersnapper was going to be his boss and someone who wasn't a traditional helicopter person, to boot.

I brought him in, told him the company appreciated all of his efforts in the past but that, like any CEO, I had to rely on the loyalty of everyone in senior management and that, all things considered, I didn't think we could work together. He wanted to know if he could get a severance package, and I said, no, that I thought the very substantial retirement he had earned and would begin collecting was fair enough.

He was pissed—certainly a natural reaction—and stormed out of my office. I later heard that he had moved to Colorado and, in one of those truly bizarre tales of good fortune, won the lottery for a couple of million dollars. He had stopped for gas at a rural station that couldn't provide full change for a fifty dollar bill, so our man took the balance in lottery tickets. The devil in me couldn't resist, and I sent him a note of congratulations, pointing out that since he would not have been in position to buy the ticket had it not been for me, perhaps he would like to split the proceeds? (My immediate staff was appalled; I thought it was pretty cool.)

A few days later, the guy called and said my letter was about the funniest thing he'd seen in years. He wasn't being snide. He also said, he'd gone to work representing a small parts manufacturer, was there any chance he might drop by and talk about opportunities? Indeed. I was pleased to give his client some business and was very pleased with the result.

That's a story with a happy ending. I have another helicopter-company employment story that does not. I discovered very quickly that we were greatly over-staffed, too many people with not much to do and the overhead was killing us. I had to cut the workforce, salaried and hourly, by 3,000. Talk about making a tough, but necessary, decision. To help as much as I could, I made public announcements in advance, alerting all businesses in Southern Arizona that some really great, highly-skilled people were about to be impacted by downsizing. I had the HR department make up lists of the skill-sets that would be available to any interested potential employers. We invited any and all businesses—including all of the major aerospace companies—to join us in a weekend job fair, which helped place a large number of those laid off. Most of the rest could take advantage of unemployment compensation for up to six months.

As I said, most. One of the company guards had taken it upon himself to pick me up at the airport whenever I flew in, day or night, even when he was not on his work shift. He had nothing in his life but his job and was happy for the opportunity to be useful. He was one of the 3,000. The day after he got his pink slip he killed himself. He had nothing in his life but his job. . . .

There is much, much more to a salesman's life than a shine on the shoes and a smile on the face.

Consultants and Agents

I got a call from a contact in Washington: the former ambassador of one of our prospective client nations wanted to be our representative. I knew the man, a big power player, and agreed to a meeting at an out-of-town hotel—my security fetish—along with one of my associates, an expert on the customer.

Power player got right to the point: "I know what you need, what you want to accomplish, and I can be of great assistance. For my part, I'll need seven percent of the deal. . . ." And he started telling me how he was going to divvy up and distribute the money—"payments" to various government officials and military officers, and I knew we were in trouble right away. He was quite relaxed, so I asked him a couple of questions, which he answered. I asked him what he knew about the competition, and he got more comfortable, and told me. I asked him about individuals and timelines, and he got really comfortable, expansive, and answered those questions and threw in a few other things as well.

After we were in my car to return to the office, my associate asked me, incredulous, "Are you going to hire that guy?" And I said, "Are you kidding? He told me everything I need to know and is so unguarded I couldn't trust him in any situation."

Since you have spent so much time and energy assembling a thoroughly professional staff, why would you hire a consultant? Because there is not enough skill, experience, or wisdom in anyone to cover all of the bases you need covered. You go to a specialist. You wouldn't have the company lawyers handle a patent infringement case, nor would you turn it over to a divorce lawyer. For example: you want to do a legislative blast—sure, you know who's in Congress and where they work, but I'll bet you don't have an up-to-date file on the names, email, and phone numbers of the people who run the local congressional offices, district by district. Somebody has the info, probably at a PR firm in Washington. So you hire them to handle the mechanics of your effort and to tell you where each member stands on the issue at hand, voting record on similar matters, and so forth, and to help you with the news media. Maybe even get a favorable editorial in a few key newspapers. They know the editors. They are specialists.

Perhaps you want to have an independent, outside study to bolster an argument you want to make to a customer. However, before you hire a consultant, be sure you have a pretty good idea what he'll find and report. The last thing you want is to get a report with the "wrong" answer. I know of one aerospace company involved in a large international procurement that commissioned a study to tell the customer which of five competing fighters would be best. The winner was the F-18. That was not the answer the sponsor—one of our competitors—was looking for, but he had to pay for it.

It is not always easy—maybe, hardly ever easy—to pick the right consultant. As in any hiring, you can't just take the candidates at their word; for some, the biggest sale they make will be themselves. They will fly to your place, look at charts, pontificate, and offer nothing of value. Some may work both sides, double agents who finagle access to sensitive competitive information, which they dangle in front of you to get more money or a contract extension, and are playing the same game with the other team.

Whether you are hiring an individual, a consulting firm, or an agency, you need to know who else they represent now, and what clients they had in the past. Are there any conflicts of interest with one of your competitors, for example? Run a background check. For international operations, check with the staff at the local American Embassy to see if they know anything, and I don't necessarily mean check with the ambassador. I mean someone with an ear closer to the ground, such as the assistant for commercial affairs. You might even hire a consultant to evaluate the candidate consultants. Overkill? When you're working a program worth billions, hardly.

Narrowing the Field

Consultants are not interchangeable, and in some parts of the world, where government operations are, well, complicated, you might want more than one consultant, perhaps even three, and you don't want two of them to know about the third. You might need an expert in the user community, another who knows the evaluators and one who can walk your team through a myriad. Suppose the nominal decision maker is the defense minister (DM). You've met him, fine fellow, very enthusiastic for your program, but is he the final approving authority? You need an adviser who knows his way around the government, who knows the strategic relationships between departments, and who knows the political relationships between key players. Maybe the defense minister and the prime minister (PM) are bitter enemies. Maybe the PM will not let the DM get credit for anything. Maybe you should be courting the PM, instead. (This is not a hypothetical; reread page 35.)

On one competition, our in-country consultant was able to convince the government selection team that they had to make a clean decision, one that did not involve bribery. He may have been helped by the fact that some people who worked for another aerospace company on an earlier competition ended up in jail. His timely exhortations worked to our advantage.

Should you jump to hire a retired military flag officer (a general or an admiral) as a consultant? It's easy to be dazzled by these really good guys. They are successful, sociable, decisive—all great attributes. However, a note of caution: for some, their first loyalty is to their service. They will give you advice and assistance, but it may be tainted by whatever is on the current chief of staff priority list. Oh, but they have contacts at the Pentagon and know all of the players. Yes, they have contacts; the friendliest among them may be the officers they helped to get promoted. Less friendly will be those officers they didn't help get promoted. There may be more of the latter than the former, and they are still in positions of authority and control. Even if your candidates are well-known and well-liked throughout their service, influence may soon enough fade away as friends retire (or political appointees are replaced), as if they came with a "use by" date on their foreheads. Keep the contracts short.

I always looked for niche players of whatever rank whose current status or contacts were largely irrelevant. I hired one retired Air Force general officer who was a leading expert on tactical air warfare. Another hire had become a "country expert" through earlier postings, who had maintained good relationships with the folks he worked with then, who were running the country now. (Yes, I know, I just suggested that government contacts were time limited and thus, for the most part, irrelevant. Usually true, but since this country was a monarchy, the folks running the country likely would be enjoying that status for many years to come.)

And there is always the standout, the exception to the "good for a few years" caution. Around 1980, we were engaged in a tight battle in Australia, our F-18 against the F-16. The F-16 team had been scoring points all over the place, their test pilot giving press conferences and getting a lot of ink. I don't remember what he was saying, exactly, but I do know that he ticked off our potential customer, the Royal Australian Air Force (RAAF)—out of bounds, if you will. This was not the way to play the game. Our team, to be frank, was not very adept at counterpunching, until someone had a bright idea, bring in Chuck Yeager as a very public consultant. Yeager, yes, the truly iconic fighter pilot (World War II ace, including one string of five kills on the same day), the first test pilot to break the sound barrier in level flight, the now-retired brigadier general. We suited him up, gave him some flying time in the F-18, and sent him off to Australia. Yeager went on the press-conference stump for two weeks, drawing reporters in with tales of his adventures. Then, he offered polite but denigrating comments on the F-16 ("No radar?") while extolling the virtues of the radar-equipped (among other attributes) F-18. From that point on, it was no contest.

Here's another great role for the right consultant. Soon after I got into the marketing business, I learned that the hardest sale to make was to my own company. I needed money to build up to, and run, a campaign; the budget-conscious company was always skeptical. Suppose I needed to have 15 million dollars in the pot—I knew that would be pretty much a show-stopper. Yes, I could juggle the numbers, tell the executives I needed three hundred thousand—an easy figure to have approved—but I would have to do that fifty times and somewhere along the way someone would catch on. Ah, but suppose I brought in an outside consultant, perhaps a well-known retired aerospace executive, to validate my approach and certify my 15 million dollar request? It always worked like a charm especially if, at times, I wrote the script (saving the consultant the time and trouble of really having to study the issue).

With some consultants, you get more than you bargain for; with others, you get less. And once in a while you may get both at the same time. A paradox? No. In my case, it was a guy named Melvin Paisley, who had recently left the position of assistant secretary of the Navy for acquisition. In 1987, my boss urged me to hire Mel—well-connected, knew the intricacies of DoD procurement. On the "less" side of the ledger, I met with him three or four times over the next year or so, and wondered why I bothered. I'd been dealing with procurement issues for almost twenty years and DoD competitions for more than half of that, and he had nothing to offer that I didn't already know.

On the "more than I bargained for" side: Without warning, I found an FBI team at my office door with a search warrant contending that—as reported in *Time* magazine—"Paisley passed 'classified and/or confidential' information to Thomas Gunn, vice president for marketing at McDonnell Aircraft Corp. in St. Louis. That information helped McDonnell Douglas formulate its plans to sell updated F-18 fighter aircraft to Switzerland and Korea. Paisley allegedly also passed along details of a competing proposal by General Dynamics to sell its F-16 fighters to the same potential buyers." For six hours teams of FBI agents searched my files, those of my secretary and some other executives, and left. I never heard from them again, but the world quickly heard about me. Through some judicial oversight, the sealed warrants implicating McDonnell Douglas and me were made public, and I became almost the only executive for a defense contractor who was, in the early press notices, publicly linked to the investigation of corruption that became known as "Ill Wind."

Ill Wind was not a mere fishing expedition, but a serious effort, backed up with solid evidence, that began while Paisley was still at the Pentagon and eventually led to the conviction of 9 government officials, 42 Washington consultants and corporate executives (including the chairman of one of our competitors), and 7 companies. One company paid some 190 million dollars in fines and fees. Paisley himself went to jail. A typical scam: in exchange for deposits, perhaps directed to a Swiss bank account, Paisley and friends in and out of government would pass along confidential competitive bid information—and then re-open the bidding so that the favored few could adjust their own bids. The term "Best and Final" took on a new meaning, with sometimes as many as three opportunities to submit a "Best and Final" offer. It confounds both sense and logic. How and when is final, final?

So, why me? Why McDonnell Douglas? The FBI had wiretaps on Paisley's phones. Paisley delighted in bragging to his friends and potential clients just how good he was, how helpful, and what wonders he wrought for me. It was 100 percent bullshit. The FBI didn't know, of course, and the news media ran with whatever they got their hands on. Think Duke University Lacrosse players.

From time to time my lawyers would ask the government, Is anything happening? Four years after my office had been searched we finally got a letter absolving me and McDonnell Douglas of any wrong-doing or involvement. *Four years.* Call it an example of our friendly federal government at work.

At times, a consultant may bring more to the table than just good advice. I've had a few who have continued to be the butt of jokes, forevermore. Helps take the tension out of a sometimes very tense business. I hired one Middle Eastern fellow who wanted an audience with the CEO. We were at the Farnborough International Air show, about 30 miles from London. The CEO's schedule was tight and the consultant was low on the priority list. The consultant understood and said, "I will pick him up at his hotel in London tomorrow morning and drive him to the air show!" This fellow drove everywhere as if he were in the middle of Paris or Cairo or Mexico City and if you have spent much time in traffic in Paris or Cairo or Mexico City, you will understand that the CEO thought he was going to die.

Another member of our consultant hall of infamy: Our representative in Holland, who called a meeting of all consultants, from all bidders in a major competition, and got agreement that whoever won, they would all share in the bonus. Huh?

A caution (one among many): Don't outsource everything or you will always be dependent upon the kindness of strangers—a recipe for failure. You must have people on staff who can do the heavy lifting, each of whom understands some important part of the business; knows the product, knows the customer, knows the acquisition rules and regulations. And be very careful about outsourcing within your own company, where some squirrels in the back room do the thinking for the front office, a new job category, perhaps, called "I do strategy for the boss." The CEO—a busy fellow with a lot on his mind—may, especially in public venues, benefit from support by a stable full of PR folk, but in my judgment, people such as the head of business development need to do their own thinking, or be given less taxing assignments. Put your qualified experts on the front line, not in the back room making viewgraphs.

Consultant Selection

The first step, of course, is to ask yourself: Do you *need* a consultant? Where does in-house capability fall short? What level of cost/compensation will the project support? Next, draft one or more statements of work and identify potential candidates. Define the consultant's role, proposed compensation, and terms of contract. Compare compensation level with that of other current or recent consultants. One consultant may not fit all needs, but all needs should be covered. As appropriate, one or more of your consulting team should:

- know the product you will be selling: capabilities, technology, track record;
- know the political environment, the key influencers, and the political process. Who is the ultimate authority? President, prime minister, king, minister of defense, and so on;
- know the customer's budgeting and procurement process;

- know the "customer" or some segment thereof: political leaders, evaluators, users (pilots, commanders, maintenance staff), influencers (media, elder statesmen);
- understand one or more of the various cultures—national, religious, tribal, individual departments of the government, regional geopolitics, public attitudes toward the military or the United States; and
- NOT be so closely connected with any customer official or other person, as to raise questions under the FCPA.

Keep a careful and thorough file of your candidates. Provide side-by-side comparisons when there are several contenders. Most likely, someone above your pay grade is going to have to validate your selection and approve the contract. Things to include:

- general background (age, education, national origin, citizenship, education, military or other government service, family and social status);
- experience, especially working with the customer (i.e., retired military) or in consulting, past and present. Are there any likely conflicts of interest?;
- Is the candidate retired (and if so, from where?), self-employed, or affiliated with a business entity (and if so, in what capacity)? Does the candidate have staff support?;
- Does the candidate envision the use of third-party consultants, subcontractors, or agents? If so, who, why, and under what guidelines and conditions?
- How did the candidate come to your attention (friends, business associates, government officials)?
- Provide the substance of at least three personal references;
- Are there any known legal or civil issues (from a records search, candidate disclosures)?
- Are there any known financial problems that could make the candidate vulnerable to various inducements that might be offered by your competition?
- Is the candidate so closely identified with a political party or lobbying group as to compromise effectiveness?

Finally—all information assembled, all references studied, some (or all) candidates interviewed—make your selection. Obtain approval to issue contract(s). Make them clear, direct, and easily understood by someone who may have marginal English language skills. Set deadlines and deliverables, such as, "Arrange a meeting with the air force chief of staff before 1 December." Also you must include a reference to, and obtain a pledge to conform with, the Foreign Corrupt Practices Act (FCPA). Put the consultant to work, and go on to win the competition.

Lessons Learned
1. Know when you are going to need a consultant, before you need a consultant.
2. Make no commitment, oral or written, or say anything that might be interpreted as a commitment, especially by someone for whom English is not the first language, until contract signing.
3. The statement of work should not be open-ended, but should define scope/duration/measurable targets.
4. Don't just turn them loose, but monitor, supervise, and require regular contact and reports.
5. Requirements will change over time. The consultant who helps you understand the culture as you prepare your campaign may not be the person who will help your customer understand you.
6. Watch for warning flags that suggest illegal or improper activity that may be just around the corner. You could be held responsible even if you were not involved in the activity, because the "flags" should have put you on notice.
    - The requested level of compensation is significantly higher than the industry standard.
    - The candidate mentions a need to make "political contributions" on your behalf.
    - The candidate requests an unusual method of payment, such as, cash only, wire transfer to third parties, direct deposit to a numbered account.

- There is anything suspect in the candidates background, qualifications, or a family or business relationship with the customer/government officials.
7. Management must understand that there will, someday, be a requirement so urgent that the selection process must be condensed. I know of one that had to be completed within 24 hours. You might go with a known entity, someone already under contract or who has worked with the company before. If you have to take on a new consultant, be as thorough as possible in the time available, and make sure the contract can be cancelled without penalty in the event of demonstrated poor performance or should any negative information surface later.

Cultural Issues

Quiz time: Would you send people out to represent your company who have no experience in the company or no schooling in the product they are expected to sell? Probably not. But all too often, companies send people who have not been culturally prepared for the country in which they are going to work. ("Well," some might say, "he's only going to England for God's sake, what kind of preparation does he need?" Ah, yes, to do business in one of the most class-conscious nations on earth. . . .)

It should be obvious that salesmen and senior executives alike need knowledge of, and sensitivity to, cultural issues in any country with which they will be doing business. Those who will live in the country obviously require in-depth knowledge. The traveling visitor can get by with token language skills and some cultural basics (although the challenge will be to keep them all straight when hitting five countries in ten days). There are many sources of good information, books, and helpful hints on the Internet. We bought language training for expatriates, language familiarization for travelers, and hired outside experts, such as the head of international programs at St. Louis University, to conduct classes.

Know that, wherever you go, you're likely being evaluated from the moment you arrive in the country by someone hired to keep a quiet eye on you. I'm not talking about industrial espionage; that's a whole different subject. (See page 90). Does surveillance seem like overkill for a simple business visit? You may think so, but your customer will not. Many cultures are keenly aware of status and will determine yours pretty quickly by how you sit (upright) and dress (upscale; look important and you will be considered important). As you check into your hotel, check your egalitarian ideas at the door. Porters carry luggage, important people do not. When waiting for an appointment, don't fidget, look at your watch, pace the floor, appear impatient, or berate the receptionist because you were on time but no one else seems to be. You are likely under observation by the receptionist and anyone who walks through the room. All may be asked for their impressions of you before your host meets you. Status-evaluation works both ways. Learn to read the signals to understand who you are dealing with.

A business meeting will get started within a few minutes in Denmark, but not for many hours in Korea or the Arab nations, where your hosts will take their time figuring you out. They view socializing, no matter how lengthy, as time invested in the business relationship. You will be offered coffee or tea, which you are to sip, slowly and in silence until your host invites conversation. Your host may or may not speak English in the meeting (even though he may speak very good English otherwise). He may do a bit of both. I had one meeting with a Saudi prince that was conducted in English, except when I made some sort of mistake, at which he shifted to Arabic for translation by his interpreter. He was sending me a message, letting me know he was important and was in charge. Let your hosts guide the conversation, don't try to change the subject and get down to business. That may have to wait for the next meeting.

In any country, clothes indeed make the man (or woman). Dressing up is better than dressing down, but it helps to know what your host will be wearing. Have your tux (or the feminine equivalent of cocktail dress or gown, as appropriate) handy when traveling in Europe. If you are a dinner guest at a private home in Holland, leave a tip for the servants; in Finland, be prepared for the pre-dinner sauna where clothing is not an issue. Everyone wears towels, unless it's an all male group, where everyone is then naked. Note that I use "dinner" in the American fashion; in some countries dinner may be lunch and the evening meal is supper.

Sharing a meal helps you get to know your companions better, and they will be taking your measure as well. (In a job interview, a meal may be part of the process. The host wants to see that the candidate can handle the niceties when dining with clients.) In some cultures or strata of society (*vide* England, noted earlier) the rules of etiquette are important, and they may vary from one place to another. In Saudi Arabia, for example, men do not eat with women except at a hotel. During my twenty-five years in and out of the country, only once did I have a meal with a Saudi wife present. In any country, wait until your host indicates it's time to be seated, don't start eating until the host does (except when there is a very senior guest of honor; the queen, for example, would start first). If there is a choice, chopsticks or knife and fork, follow the lead of the host. If he picks up the silverware, he is being polite. If he uses the chopsticks, he may be testing you. Learn how to use them before you need them; practice picking up peanuts. Should you order pork if it's on the menu in an Israeli restaurant? No. If you are in China, leave something on the plate; if you finish everything, your host will think you didn't get enough to eat and are still hungry. In Japan, slurping your noodles is a compliment to the chef; in France, asking for ketchup is an insult to the chef.

Such cautions work both ways; I remember well a visitor from Switzerland who devoured the flowers in the table centerpiece, much to the amusement of the restaurant staff; he came from a country where they eat flowers. Or the Korean Air Force officer on his first international trip, who gamely tried to chew his way through a Maine lobster, shell and all. He came from a country where they don't eat lobsters.

Be careful lest your own efforts to fit in and be gracious go awry. John McDonnell and I were at a very important dinner in China, when I decided that John should offer a toast, in Chinese, of course, to the host's daughter; something about, what a pleasure it was to share a meal with such an attractive young woman. I had it written out phonetically and a Chinese general helped John with the pronunciation. But Chinese is a tonal language; you can make the right sound but at the wrong pitch, so to speak. John stood up, delivered the toast, and everyone started tittering in a polite

Chinese manner. After dinner, a puzzled John asked someone, "What did I say?" Someone replied, "She looks like a horse." That's the last time I tried that trick.

Wherever you come from and wherever you're going, know the rules of the game before you decide to play. You will hear over and over again the do's and don'ts: In the Arab world, never show the bottom of your shoe; never shake hands left handed; shake hands with a woman only if she offers hers first (and not in a public place); never look a woman directly in the eye except perhaps for about a second to acknowledge a greeting; never be alone with an Arab woman unless you are in a situation where class standing and dignity don't count.

If you try to enter Saudi Arabia with pornography, you will be turned back at the airport and put on the next available flight out. The merely (in our judgment) risqué—racy romance novels—are not welcome. If you have a *Playboy* magazine, it will be confiscated but you may be permitted to remain. The fate of the magazine will be unknown.

United Arab Emerites (UAE) and Kuwait are more tolerant of Western manners and morals, but be not disabused. Treat the apparent relaxation as a concession to the needs of business. Underneath the surface, the rules are still there. Case in point: a group of happy salesmen had a few beers and spent the evening cavorting with some local women in a hotel swimming pool in Abu Dhabi. Flirting, playing tag. Everything was caught on the hotel security cameras, innocent enough in our eyes, but so outrageous to the hotel management that they closed the swimming pool—for a year. Men and women do not swim together. In truth, in some Muslim countries, when women go swimming at all, it is usually with their children and they are fully clothed from head to toe.

A full-time posting to Saudi will be challenging. About 100,000 Westerners live in the desert kingdom, most in the capital city Riyadh and most of those live in walled compounds, within which they can do much as they would at home. Any given compound may house as many as 1,000 people and have a shop, restaurant, movie theater, hairdresser, sports complex, health center, swimming pool, school for small children, and shuttle buses standing by to take women outside the walls for shopping. This is a major source of recreation and amusement for which they must be fully dressed in proper Saudi female kit. The religious police do tend to be more forgiving of minor infractions by Westerners, like a short skirt—that is, an inch or so shorter than the prescribed ground-length.

Inside the walls, the women can run around in shorts, sandals, and tennis outfits. The single men chase the single women (usually, nurses; it takes a fair number of nurses to care for a population of 100,000). Since there is not much else to do, they have a lot of parties where, frequently, alcohol is served.

Please understand, alcohol is totally banned in Saudi. But, as America learned in the era of Prohibition, if Americans want to drink, they will find a way. "Bathtub gin" is not a myth, but the name of a recipe. In those Muslim countries that have the strict no-alcohol rule, some Westerners brew their own. Call it an open secret, at least in Riyadh. The government knows they do it and they know that the government knows it but they do it anyway.

However, should an alcohol-consuming expatriate get cross-threaded with a neighbor or the government for any reason, he is likely to be charged as a user and given 24 hours to leave the country. It's easy to understand the chain of events if an angry neighbor turned him in, but absent a complaint, how would the government know, specifically, who drinks and who doesn't? Just assume that anyone who enters your home to cook or clean or deliver—people who are most likely part of the imported labor force—depend on the goodwill of the government to remain. The "goodwill" may be preserved by passing along, ah, certain information.

The Muslims worship five times a day (allow time in your work schedule and business entertainment for the mid-afternoon and evening prayers). Personally, you will find limited opportunities to worship in Saudi. McDonnell Douglas used to hire a priest from Kuwait to fly into Riyadh about once a month to say Mass in an employee's living room. Protestants often relied on lay leaders or took turns reading scripture.

Study though you might, there are some cultural missteps for which you really can't be prepared such as the actions of some really dumb (or merely insensitive, but I think "dumb" is a better description) representatives of your company. Here are some examples.

- After a frustrating business meeting in Paris, a senior executive said to me, "These f___ing Koreans, why don't they bother to learn English?" He was speaking privately, but do you think that his attitude didn't show through to the customer? Not the sort of person you want to have in sales. Or maybe in any job.
- We wanted to sell some airplanes to Morocco, then in the buying mode for some fighters. The Moroccans were interested, and the ruling sheikh wanted to meet our president. Well, it was at the time of the Paris Air Show and the sheikh offered to send his personal airplane to Paris, to fly the president down for a visit. The president couldn't be bothered; it was not as if he was overwhelmed with show business because he left for home a day early. The sheikh never forgot the snub. And we never got any orders from Morocco. Not just then, but ever.
- In America, it's common for a salesman to flourish a big-name designer fountain pen for signing a contract and then tell the customer to "Keep it, it's yours." (In my experience, the pen may be counterfeit but the gesture is none the less, ah, heartfelt.) In some parts of the world, such forward behavior counts as an insult. I know of one prime minister who was so outraged at the impudence that he called the company president to complain. I don't think the foolish salesman remained in that assignment.
- And what can you do with the guy who went to Greece to pitch a sale, turned on the projector and lo, the first slide said, "McDonnell Douglas is pleased to offer Harpoon to our Turkish friends."
- Douglas lost what their president thought was a sure thing—an MD-11 sale to Singapore. The country bought 747s instead. The president of Douglas thought he'd been double-crossed and wrote a nasty letter to Prime Minister Lee Kuan Yew accusing someone of shenanigans, a hint of payoffs. I happened to be in Singapore at the time, trying to sell some military aircraft, and was called on the carpet by an angry PM, who said he would never forget or forgive. I asked John McDonnell to fly in and join me for some damage control by having a private meeting with the PM. This helped a bit, but not very much. Singapore ignored our offers and ordered seventy

F-16s. It was fifteen years before they bought any McDonnell Douglas products. Thus, because one part of the company—Douglas Aircraft—was not aware of (or paying attention to) what another part was up to and because a senior executive insulted a head of state, we may have lost a couple of billion dollars in sales.

Much of this just comes down to common sense and courtesy. Keep your promises, and stay in touch. The industrial participation arrangement on one European sale included licensing a domestic company to produce a very sophisticated precision measuring system. A couple of years later, one of our team was visiting an acquisition official at the Ministry of Defense, to explore a new opportunity, and before he had barely started, perhaps hadn't gotten beyond "Good Afternoon," the official went ballistic. The offset deal had cratered. We hadn't lived up to our promises. Hadn't we heard about the lawsuits? Well, no, but we sure would look into it.

If this was a simple contract failure, perhaps caused by poor communications and misunderstandings, why was this fellow's anger so over-the-top? As it turned out, in at least three visits to Parliament, the defense minister had bragged about snagging this particular offset deal, high prestige for the nation, a real coup. Then the political opposition learned of the lawsuits, charged some form of perfidy, and the minister was greatly embarrassed.

We pulled out all stops, re-engaged with the local company—and their lawyers, uncovered the disconnects and got both parties, us and them, fully involved. Both parties admitted to some part of the blame; both agreed to charge ahead, refreshed. The defense minister could share in the credit for the turnaround and became our new best friend.

Finally, not all cultural issues are overseas. You must educate your senior leadership as to the foibles or peculiarities of any customer. You never want the company president make casual conversation with, say, an Army general, by asking, "What are all those decorations on your costume?" (I am not making this up.) Also, each of the armed services has a heritage. It helps the conversation if you know something about each. I don't mean bone up on the War of 1812 or Iwo Jima, but have some idea of what they have done with airplanes, especially in recent years. The Army is damn proud that their Apaches took out more than 500 Iraqi tanks in the first Gulf War. You get the idea.

On Dealing with Congress

Most of our domestic sales were to an arm of the federal government, principally the armed forces. You might think that purchasing decisions or creating military R&D programs are the province of some general or admiral, or the secretary of a department, following the recommendations of teams of experts. Well, perhaps, but except for company-financed programs, the money comes from only one source, Congress.

Congress exercises total power in the realm, for good or ill. If Congress wants to kill a program, it can. If Congress wants to support a program that the military doesn't want, the military gets the program anyway.

If you are an up-and-comer, just out of school, or in your first marketing job, the following few pages may offer the best advice you will get, anywhere, on how to succeed in selling to the government. If you are an old hand, you may enjoy a review of a few trials and triumphs.

We start with the basics. It may seem obvious, but is nonetheless important to note, that members of Congress hold their jobs at the pleasure of their constituents. If they do a good job, look good doing it, show some spunk (particularly if most constituents agree with the position espoused), demonstrate enough party solidarity to garner choice committee assignments, and provide good service to the hometown folk, they will likely be rewarded with new terms *seriatim*. If they stay in Congress long enough to become an icon—I'm sure you can think of a few examples—local pride will keep them in office, possibly for the legislative equivalent of forever.

To be reelected, members need visibility. So do opponents hoping to unseat a member. Unless one is an icon, visibility costs money. Money has to come from somewhere. To be brutally frank, the voters in any district, themselves, do not have enough money to support a candidacy in today's media-driven campaign frenzies. Money comes from many sources. We started a McDonnell Douglas Political Action Committee (MDC-PAC), which, guided by a steering committee, allocated appropriate company contributions to selected members of Congress. We encouraged our executives to support campaigns. Money comes from other political action committees, businesses, unions, corporations, and committed activists, many if not most of whom have no tangible ties to the member's district. It is not uncommon that they may have donated equally to candidates on both sides of the election—one of whom will win.

Do not be confused: money does not buy unconditional support for your programs, unless, perhaps, your Congressional representative is a crook. Money buys access, granting you the opportunity to have a conversation on the telephone, or a meeting. When a call, email, letter, or memo arrives at a member's office, the staff runs a quick screen (very quick, if it's a phone call flagged as "urgent"). If the supplicant is a donor, close friend, or colleague, the call goes through, and the meeting may be arranged with more dispatch than for the typical supplicant or constituent. Be not outraged; it's the way much of the world works.

However, access does not guarantee anything more. From that point, the donor, friend, or colleague has to make a good case to get support. "This program will bring a thousand new jobs to the district," is a good start. "We've done a survey of your constituents and they are overwhelmingly in favor," might also work. However, "This program is vital to national security" may or may not have some impact unless you really lay out the case. Of course, the effect will be greater if combined with one of the other approaches, but if the program is a real dog, of value to no one other than the petitioner, access gains you nothing, although you may not know it until the floor vote is counted.

The individual members of Congress are bombarded from all sides and are targets of every special interest group (and every one of your competitors). You need to know the gatekeepers and have them know you. The chief of staff is a key, but not exclusive, player. A solid contact with almost anyone on the personal staff is useful, and don't overlook the folks in the local district offices, especially in those districts where you have facilities, major suppliers, or other operations. Through them, the members can be kept informed of your programs and progress, employment, or other legislative issues. Assisted by a top-flight D.C. PR firm, my own staff would assemble briefing books, tailored for each district, with data and photos of our facilities and our programs, both ongoing and those in the works. We prepared a more general briefing book for members who had prior military service but whose districts did not include any of our facilities. We summarized important issues, Congressional or international, and included a "white paper" or Q&A sheet that would quickly bring any member up to speed.

To ensure that individual briefings are on target, you might assess the depth of the member's technical knowledge. A member of Congress who wonders why you don't mount a couple of the engines backward to help stop the plane faster is not a good candidate for a detailed analyses of specifications and cost. (Again: I exaggerate, but not much.) Concentrate on everyone's favorite topic—jobs.

At least twice a year (but not in the middle of an election campaign) we would bring the members out for a local plant visit, meet-and-greet some voters, perhaps lunch with the executives and union officials. Once a year, we would try to give them a ride in one (or more) of our airplanes. This was not sight-seeing, but a solid demonstration of capabilities.

We asked our union leaders and major suppliers to visit members in their D.C. offices, partly courtesy, partly pleading. We encouraged our international customers to invite members for a working visit and some friendly lobbying. We also met with the editorial boards of national and regional papers, making sure that they were aware of whatever issues we were facing, arming them with the facts they might use in writing editorials, and had friends write supportive OP-ED pieces or letters to the editors.

There is a banquet of some sort almost any week in D.C.—arts organizations, charities, associations. Members want to be seen as supportive, so we bought tables and invited members, key staffers, and their wives, to come as our guests. (Note well: the rules have changed, and this sort of hospitality would not be possible today.)

The members vote, but the things they vote on and the reports they argue about are drafted by staff in the various committees. These are the first people who need to be convinced of your arguments and of the merits of your program. They help create policy and quantify costs. You would naturally focus on the Armed Services and Appropriations Committees, but don't ignore the House Science and Technology Committee and the Senate Committee on Commerce, Science and Transportation, which oversee space and aviation.

We used PR consultants and hired former members as advisers, to help us know the real centers of influence—members, specific staffers—to tell us how best to state a position or frame an argument in a succinct, reasonable, and non-argumentative manner.

While all of the above will contribute to access, true personal relationships, which take more effort, over some period of time, are likely to prove of greater value. The business school buzzword is "networking," creating and maintaining a wide circle of contacts who might be helpful with one or another problem or project many years out in the future. I suppose that sounds a bit devious, evoking the marriage of a huge Rolodex and a calendar noting birthdays and anniversaries to be acknowledged, but I prefer to call it gaining and maintaining friendships. Properly done, it is invaluable in life in general, and adding to success in business is, well, a plus.

For example, shortly after I took my job at the FTC, another lawyer in the office mentioned that he was "going over to Joe Addabbo's office to stuff envelopes." That sparked my curiosity; I knew that Addabbo was a Congressman from Ozone Park, Queens, New York, and one of my close friends from law school, Louis Desena, was from Queens. So I called Louis, and he said, "Joe Addabbo is my law partner!" I joined the envelope stuffing and Addabbo bought me a beer (a man who buys you a beer can be a friend for life). A few years later, when I was head of the Washington office, I heard that Joe Addabbo was having a big fund raiser in his home district and, minding my manners, flew up to make a contribution. (I also dropped in on Louis; the Congressman's district office was on the first floor, the law firm was on the second floor. I doubt you could have that sort of arrangement today.) Louis, Addabbo, and I had lunch. I paid. Fast forward a few years, our F-18 program was in trouble, the House Appropriations Committee Defense Subcommittee—chaired by Addabbo—wanted to cancel the program, I called Addabbo. He was a vocal critic of Reagan's defense buildup. I made my case. He changed his vote and got a few other members to do the same. The program survived. Point made?

Friendships work, of course, in many ways. While I was with the Senate, I also became close to Dan Tate, a whiz at dealing with the Senate Finance and Ways and Means Committees, and Ken Duberstein, a staffer for New York Senator Jacob Javits who became an expert in the arcane workings of Congress and the Executive Branch. Over the years, along with Washington PR-guru Gerry Cassidy, they were my go-to guys. These were men I could count on for good advice and great assistance when dealing with Congress, the administration, the Washington press corps, and grass-roots support organizations. I don't want to slight any of the hordes of consultants and agents I worked with over the years, but these three were, well, the best.

And they were good friends, as well. In 1978, Tate was deputy director of Carter's congressional liaison office and recommended me for a White House job. In 1980, I helped Duberstein write a resume that would get him a job with the White House (he later became Reagan's chief of staff and is today the lead director of my last employer, the Boeing Company. I put in a good word).

Here's a brief but significant digression about that job with President Carter and why I didn't take it. When Dan Tate reached me in St. Louis, he got right to the point. "Tom, the president is slipping in the polls, we need to begin working on the re-election campaign right now, and, well, would you be interested in coming back and working with us?"

I let the Washington office know that I was coming to town, booked a flight on the next plane to D.C., checked into a room at the Mayflower and went straight to an arranged meeting with Carter's chief of staff, Hamilton Jordan (pronounced in the Southern fashion "Jerden"). We had a long talk, especially touching on my campaign experience with McClellan, and since I was clearly interested, he said, "Let's go see the president."

Carter was most gracious. Whatever was contributing to his slippage, it wasn't bad manners. He said, he understood that I might be joining the staff, coming "with some very strong recommendations," and hoped that I would do so. Even for an arrogant young guy like me, standing in the Oval Office and being offered a job by the president was a transcendent experience. But I had a partner in my life, and had to talk it over with my wife and said so, which certainly didn't hurt my standing with Carter.

Ham took me back to his office for some mild arm-twisting. I definitely was interested, but I told him that I would indeed have to bring Kate into the discussion and would get back to him as soon as possible.

It was now well into the evening, and by the time I got back to the Mayflower the word was out that I may be going to work for the president. Hard to keep secrets in Washington! I discovered a bunch of my friends and colleagues had gathered in the bar, throwing a party on my tab. Well, I joined in, big time—I admit today that I was a heavy drinker then.

About 2 AM, I left the party to go to the men's room and the clerk at the front desk caught me. "Mr. Gunn," he said, "we have an emergency call from your wife, you have to call her right away." I grabbed the nearest house phone and called. Kate said our seventeen-month-old daughter Megan had been diagnosed with a very serious illness. I said, very serious, what does that mean? And she said, "Her life expectancy is two years."

In a daze, I staggered out the door, went walking around the streets, trying to sober up, trying to think. St. Matthew's Cathedral was a few blocks away from the hotel, and I pounded on the door of the Parish House until a priest came. I handed him a wad of bills, all the cash I had in my pocket—maybe 300 dollars—and sobbed, "Please, say a bunch of masses for my daughter Megan, we just found out she has a terminal disease." And at that moment I took the pledge, and I've not had a drop to drink since that day. My personal priorities changed, and any thought of moving job and home to Washington evaporated, and Megan thrives some 30 years later.

CHAPTER 3

The Business of Selling

In this chapter we'll take a look at the general subject of "sales," (with a largely international focus), the value of air shows, and offer thoughts about competitors, turning competitors into partners, espionage, the ins-and-outs of industrial participation commonly called "offset," and run through a typical DoD Request for Proposal (RFP).

Sales

As the New Business Activity matured, I changed the name of the sales and marketing department to Program Development (soon modified to reflect a broader focus, Business Development). Why? I learned that the folks handling security clearances thought that sales and marketing guys just naturally talked too much and couldn't keep any secrets; as a result, we were being denied access to important program information.

It may seem too easy: change a name, change the clearances, but it worked. I did not, however, stop using the word "sales." Sales manager, sales force, tells the customer who you are and helps remind the members of the team what it is they are supposed to be doing.

The ideal customers, of course, are those who buy your product, use it with pleasure, and never quibble about the price. I'm not sure if I've ever met one of those, but some I've worked with are really impressive.

Not to mention any names (I am in the aerospace, not bridge-burning, business) but the best international customers tend to be disciplined, with an organized selection process, honest to a fault, and pay their bills on time. They are predictable, in a business sense. It may take a long time to come to a decision, but when they get there, they stick to it. Other countries are more like us, not so organized but decent fellows (although perhaps too influenced by indigenous aerospace companies competing for the business).

Next down the line are the economically successful but unpredictable. They push for every advantage (not necessarily a bad thing, except when certain under-the-table investments are suggested). Be wary of customers who might want to pull you into a deal, say a contract for 100 airplanes with first-year deliveries of two or three and who, after they have plundered your technology, might cancel the contract. Next, watch for customers in countries with unstable governments who offer unstable agreements. Finally, you have the window shoppers and tire kickers. You wonder why you bother but once in a while you strike gold. More often you strike out.

If left to their own devises, some in management might be reluctant to invest large sums of money in pursuing international sales. Once, I made a pitch: to compete for the business of a new overseas customer, we had to invest some 5 million dollars. One of the corporate executives said, "Why would you do that?" and I said, somewhat loftily, "To make the sale." He said, "Tom, international only represents about thirteen percent of sales, that's all out of proportion." I let a bit of temper get in the way, looked him dead in the eye in front of a room full of very senior people, and said, "Are you telling me that because international only represents thirteen percent of sales right now, that you're happy with that? You never want to go above thirteen percent?" I gave him a little lecture. "Here's a lesson I learned from Mr. Mac. It costs so damn much money to build these airplanes that we can't possibly amortize the cost with just domestic sales. The Air Force or whoever isn't going to buy enough of them fast enough. The company seems to have forgotten that Mr. Mac, God rest his soul, had an international target of twenty-eight percent, where sales were nicely balanced and the company made money." And the now-chastised executive said, "Oh." I got the five million.

We sold to stable countries but not necessarily stable governments. In the parliamentary system the ruling party can be dumped at any time; we used to say (only half in jest) that we would see two changes in government before completing a sale. I say only half in jest because we worked on a sale through more than one government in Switzerland, Spain, and the U.K. Clearly, you need to make your pitch not only to the government of today but the likely government of tomorrow. Be a friend to all. But especially, be a friend to the military, unlikely to be changed much in any change of government, and seek out professional bureaucrats who know their way around the system and will endure.

In making your pitch, keep in mind that what you say may not always be what you are thinking; what the listeners hear may be something yet again different, colored by their own assumptions and past experiences. Almost like the party "telephone" game with only two players. Message sent is not necessarily message received. Much of the time many of us don't fully listen. We're distracted, worried about something, or mentally leaping ahead to what we are going to say next, which may have little to do with what we're hearing.

Treat a sales call like a job interview. You want to put your best foot forward, with economy and clarity, have friendly but not over-animated interaction, and have an agenda, an outline, and a set of talking points firmly in mind. I know, this seems too basic to have to mention, but after you've sat in on a few wasted sales calls, you'll understand. Trust me: have an agenda, an outline, and talking points. You may or may not be able to follow all, and you don't want to sound like you've memorized a script, but you must know what it is you want to talk about, in what order, and to what purpose.

Let your customers know that you've been listening to them. Repeat a key phrase, or re-word a comment. "Now, let me be sure I've got this, you want help with financing?" Ask questions to draw the customers out, not flat statements to shut them down. "What options did you have in mind?" rather than "We sell airplanes, we don't rent them."

For the record, there are no really bad customers. There are just different customers. You must understand the differences. If you don't like your customer, you might as well stay home. I liked them all, every one, even the most troublesome, because I was dealing with people, not governments.

A suggestion: It helps to know where your customer's key players fall on the scale. Users, evaluators, decision makers, you want to do things to make them feel good. Borrow a page from some "Bartender's Playbook" that advises, "Tips are enhanced if your patrons think they are really liked." Everyone has an ego—some egos are bigger than others and need regular feeding. Use your imagination. Invite them to something not prohibited by regulations or proscribed because you are in the middle of a competition. See if some college or university might want to grant an honorary degree. Arrange an invitation to address a trade group or civic organization, where you get to make the introduction. Hold a charity tournament; it doesn't hurt if it is the favorite charity of the chief of staff's wife. Invite her to play. With the approval of DoD, invite an international customer to witness a military exercise where your products will be involved. This is very effective, especially with relatively low-level people who never get a chance to get out of the office, let alone out of the country.

If you are satisfied that the opportunity is real, the customer is facing a real threat and has a real need, you need to find out if they have the money. Our sales were so big, we knew that only countries with a certain level of gross national product, those with an established middle class, could afford them. We didn't much bother with expressions of interest from the Maldives. Our sales were so big that the customer—a government—would most certainly have to commit the money in advance. If I had any doubts, I insisted, before we would answer the RFP, that they show me the money. Oh, they might say, our budgets are classified. Of course, I would say, but before the State Department will give me a license to reveal technology, I have to assure them that the deal is real.

You have to know where your payments will come from, and it's not always easy to tell. Countries may have political budgets that they show the people, hidden budgets that reflect what they really expect to spend, and a secret pot of money that's held by the ruling family which may be used to support certain programs, such as national defense, and they don't want the neighbors to know what they're spending. It's hard to follow that money trail. If I think I see a problem and someone in the finance ministry tells me, "Don't worry, the money will be there," I worry. If I hear that from the crown prince, I relax.

Money in the budget is one thing, how that money is going to be doled out is another. You must have a payment plan, with a certain amount to get started and so much for development and procurement (which may be spread out over many years) and post-delivery support.

There have been times when "committed" funds would slip, as in the case of the Saudis in the mid-1990s, who were faced (for a short time) with falling oil prices. We knew they were good for the money, so we juggled some deliveries and there was no crisis (see page 139).

Here's a general but greatly simplified look at the steps along the way to bringing forth a military budget. You need to have someone on your team who really understands these things.

Budgeting is fairly straightforward in our DoD, sort of a series of waterfalls pouring into ever-enlarging pools, with the size of the pools more or less established in advance as budget targets. The users and operators sit down and tell their service what they want and need. The service sorts it out and sits down with DoD to submit their near term and long term budgets—including personnel, operations and maintenance, and the thing you're most interested in, procurement. DoD tries to keep some balance, wraps in deep-think on "the threat," what new technology needs to be explored, and issues a fiscal year budget and a rolling Future Years Defense Plan (FYDP, pronounced "fiddup," which, for you old-timers, used to be called the "Five Year Defense Plan") taking a look five to six years out. The DoD budget is folded in with requests from all the federal departments—there may be adjustments imposed by the Office of Management and the Budget (OMB)—and the result is submitted to Congress. The appropriate committees pull out the appropriate sections and the legislative process begins. Bear in mind that there are at least four discrete points before or after which you can try to influence the procurement: users to service chiefs, service chiefs to DoD, DoD to the administration, the administration to Congress. Plan ahead and start early.

The legislative process is similar in most other constitutional democracies (France, India, Finland, Israel), and is not much different in strong parliamentary republics (Singapore, Germany, Pakistan) with elected Presidents and Prime Ministers, or in constitutional monarchies (U.K., Spain, Japan) where the monarch is head of state but the Prime Minister, who got the job directly or indirectly through an election, is the head of government. In countries where there may be an all-powerful reigning monarch, he may keep all power unto himself or delegate military purchase decisions to his finance minister, defense minister, or cousin. You need to know who is which and has what authority.

There are, of course, other forms of government. You may or may not ever do business with the Democratic People's Republic of Korea or the Socialist Republic of Vietnam. But you never know. Plan ahead and start early. You would likely need a lot of permissions.

The company wants the highest reasonable price (there is a lot of unallocated overhead and restive stockholders; the board of directors needs to show profit), the customer wants an unreasonably low price, and marketing is in the middle. I've said it more than once: your hardest sales will often be to your own company. We had trouble selling the idea of a "win price" (that is, the price it would take to win the contract). We might as well have been talking a foreign language. They would say, "We know what it costs, we know what we need to make a profit, what else is there?" "There" is the price at which the customer will buy your product rather than someone else's, and it most likely will be somewhere in between the poles. So we would agree on a price, but when it came to the Best and Final Offer (BAFO), if price was to be a consideration, finance would balk. "The price is the price." Well, yes and no. If by this point you know for certain that it's more than the customer can (or is willing to) pay, it's no-win price. However—a big "however"—suppose that making this sale means keeping tens of thousands of people gainfully employed as the production line remains open, ready to absorb new business, but you can only do the deal at zero net? My vote: mind the cars in the parking lot and go on the road to look for new opportunities.

Be very careful if, and when, you ever discuss pricing with the customer before you have submitted the proposal. If you suggest a price, no matter how casual the conversation, you might set up false expectations. Low-ball the numbers and you will lose credibility when the proposal is opened, but at the high end (which may be more honest) you could trigger sticker shock before the customer can even read the proposal and study the explanation.

Pricing is sensitive to many things, but one above all: time. You must have a "this pricing is good until midnight on a certain date" clause to protect you against vagaries in your costs. A world crisis could spike the price of Titanium by 400 percent or a new labor contract could impact production. Your management will be pushing for a price increase and your customer will think you're pulling a fast one. Keep smiling.

When your competition is a nationalized or government-subsidized company, they can price their products pretty much wherever they want, they don't need to cover costs, overhead, and show a profit. But your customer should be interested in more than just a low price. You need to counter with the "ilities": capability, durability, reliability, maintainability, and emphasize your reputation for follow-up dependability.

Field Offices

When do you establish an international field office? You open an office when you have made a sale. It's too hard to close one down if you lose a competition because you'll likely be stuck with a five year lease. After a sale, you need to have an in-country presence to coordinate support and handle problems. There is only marginal additional cost to add a salesperson to keep up with the customer. Many of our active field offices were established back in the days of the F-4 Phantom or some commercial aircraft, and over the years assembled a great deal of customer knowledge and developed a great deal of good will within the community.

I have a very precise vision of what an international sales office should and should not be. "Should" should be obvious; an office staffed with a professional salesperson or two and modest administrative staff, but there are a lot of "should not be" offices around the world trying to be everything to everyone. I've seen more than one major-city sales office set up with a large staff and a very senior non-salesperson in charge, with a charter to do "country analysis." And oh by the way, when you have a big office you raise big expectations. Is there a charity drive or a new museum about to open? You'll be invited to make a "suggested" donation commensurate with the apparent size of your ambition, perhaps around 100,000 dollars.

The money for all of this comes out of the marketing budget; I vote to put it to good use—selling. The State Department and the CIA already do a good job of country analysis, information that is publicly available and can be easily expanded, as needed, to include data of interest that wasn't on the government's checklist.

My rule of thumb is to start small until you see what you need and expand if and as necessary. Our office in Tel Aviv started with one man, then we added a couple more and then, as we got to know the customer and because we were right across the street from the resource-limited military headquarters, I expanded to add a couple of conference rooms for use by the military staff. They came in every day to use our rooms, and we were hands-off. I'm quite certain they didn't talk about classified or sensitive matters but had routine staff meetings, briefings for out-of-town visitors, and cigar-smoking sessions with my staff after working hours.

About Air Shows

Almost 150 communities in the United States host an annual "air show," bringing in aerobatic performing acts and ground displays of military and historic aircraft. But don't get confused: those are entertainment, not marketing, shows, even though local aerospace firms may sponsor a modest good-neighbor presence. Those shows give the folks something to do one weekend a year and help get youngsters interested in aviation. In truth, the military aerobatic performing teams, such as the Navy's Blue Angels, demonstrating precision flying in our F-18 Super Hornet, are funded through recruiting budgets and are focused on getting qualified students into pilot training programs.

To the aerospace industry, an air show proper is a for-the-trade exposition, combining flying demonstrations of products for sale with static displays, exhibits, press conferences, and venues for private meetings. The granddaddy of them all is the Paris Air Show, held every two years, which brings out two thousand exhibitors and half-a-million attendees. Half of those are trade visitors from some eighty-eight countries, half are members of the general public. They are allowed in the last three days of the show when most of the exhibitors shift to skeleton crews because, well, the general public doesn't buy many airplanes but the rules of the show require that all exhibits be staffed.

Though such a major venue may seem important, at my end of the business, military aircraft sales, it was not much. Nor were any of the twenty-odd aerospace trade shows held every year worldwide. Selling military aircraft is an on-the-ground enterprise, and in any given year, there may only be one or two potential customers. Why spend millions of dollars to show off your product to a very local general public and to your competitors? (A note: espionage is not limited to governments. You no doubt will try to bullet-proof your exhibits from reverse-engineering. Your competitor will try to get a quick peek, and bribe the freight forwarder or whatever guards you might hire. Be prepared. If you want to do likewise, be forewarned: it's probably against the law in some country, probably your own.)

But, millions of dollars? That's at just *one* show. Plus, you must invest several months to prepare the messages, prep senior executives (yes, practice makes perfect), and to get to some venues you will be trapped for as long as twenty-four hours in two or three or four airplanes, any one of which may have captured your luggage. Plan ahead and travel light. At show time, you will spend three to four hours a day commuting from central city hotels to the show site, more if there is an accident (this over a route that might take twenty to thirty minutes each way at midnight). Did I hear a suggestion, "Stay in a hotel near the show"? Some companies have tried that, and some companies always station low-level staff in the immediate neighborhood, but all senior folk and anyone in sales will be spending every evening in the center of town. The receptions, the parties, the socializing are in Paris, London, Singapore, not the outlying airports at Le Bourget, Farnborough, or Changi where the shows are staged. Building relationships is more important than showing product.

A big air show does give senior management a chance to meet important customers over lunch, dinner, at an invitation-only reception. It also grants a sense of prestige to smaller customers, invited into your inner sanctum, the so-called "chalet," your pro-temps, entry-by-invitation restaurant at the show site that will be open for only four or five days and on which you will have spent more money than most people spend in building a home. Food, beverages, and serving staff are extra. Budget one hundred dollars a head for lunch. Having a public demonstration flight at the show? At the close of the day, your airplane is available for a private demonstration flight for a potential customer. Take advantage; this may be the best reason for being there.

A big air show also gives the aerospace trade press a chance to gather nuggets and thereby multiply your exposure, and part of your job is to make sure that interesting nuggets are waiting to be found. I'm not just talking about company hand-outs; a journalist doesn't need to make the trip to get something he can find online, but he needs to file some good copy with his employer to justify the expense. Help him out and you have a friend for life (or at least for a couple of weeks). Whatever, it's worth the effort. Far more people read the aerospace press than attend any, or all, air shows. If you have a really good story to tell, you'll make mainstream media. A positive profile in the *Wall Street Journal* would go a long way to offsetting some of the cost of the show, although it's more likely to impact your stock price than your sales. That's not at all bad.

At the more personal level, an air show may provide your spouse with an opportunity to get out of the house and see what it is that you do for a living. Even if you are not a very senior person whose spouses are part of the official program and thus sponsored, the nominal cost of taking your spouse is airfare. The hotel and most miscellaneous expenses are covered. (NOTE: I said, "nominal cost." That does not include the time your spouse might spend a month's worth of your salary on a must-have antique.)

For commercial aircraft sales, exhibiting—and the attendant worldwide press exposure—*is* important. It's a far easier method for reaching the world's two-hundred-plus airlines and aircraft leasing companies than flying from one to another like an old-time door-to-door salesman. Piece-part and systems manufacturers also benefit by the exposure to airframers.

Whether your mode is commercial or military, there may, from time to time, be transient political considerations. If you want to do business in some off-circuit country that suddenly decides to host an air show—Indonesia, for example, which held two in the past twenty years—you would be well advised to sign on. This has little to do with selling; it is buying a ticket that allows you to come back for a sales visit.

While some of the reasons for exhibiting are logical, the profusion of major shows is not. Some shows are annual, while the larger venues tend to be bi-annual: Paris and some others are held every odd-numbered year, while Singapore and other countries hold shows every even-numbered year. We are not working in an industry, such as consumer electronics, where a new model comes out every couple of months. It takes a long time, measured in years, to bring almost any aerospace product to market; not much changes year-to-year. I think that shows should be on five-year cycles.

I am not alone in this. Over the years, there has been industry talk about arranging a boycott of sorts, cutting back on participation and forcing the air show industry to pay attention. One year, one of the second-tier manufacturers skipped one of the big events, trying to send a message. The message received was that they were in trouble, couldn't afford the cost, and were up for sale, whatever. They were back at the next show.

I will admit that my company almost backed out of the 1991 Paris Air Show. We did not plan to exhibit, perhaps have a few of our people attend and to just wander around the exhibits, collecting brochures. However, Saddam Hussein invaded Kuwait in August 1990, and by the following January the United States was at war. The Department of Defense asked us to change our mind. It wanted to capitalize on the exemplary performance of American equipment in the war. However, we were then so late in making our reservation for exhibit and chalet space—well after the nominal deadline—that we ended up in the equivalent of a double-wide manufactured home stuck way out at the end of the line. The only people who could find us were our close customers, because we made sure they knew where we were. That was not all bad, it cut down on the tire kickers who were just looking for free drinks and allowed us to spend more quality time with the people who mattered.

There was another unexpected benefit. DoD had provided a couple of European-based F-15s and a Harrier for display, and we wanted to exhibit the Apache AH-64 helicopter, one of the aerial stars, credited with taking out Iraqi communications the first night of the war. (We ran an ad in the trades, "No matter what your mom said . . . there is a reason to be afraid of the dark.") Not to mention the Apache's role in destroying some 500 Iraqi tanks. However, there was no conveniently-available Apache. Most were still in the Gulf region or locked in the supply pipeline. We thought about taking one off the production line and prepping it for public display, but this was a thought that did not sit well with the Army, which wanted every flyable bird in the inventory. So we paid for a parking space on the flight line, but instead of an Apache, we set up a large placard with a picture of an Apache, headlined "Called to Duty" and describing key features. It cost us two dollars and we saved about 1 million—what it would have cost to indemnify the Army and prep and transport an Apache. We got more good press than if one had been on display.

As a footnote of sorts, I should note that, even though the Apache had done a bang-up, knock-down job in the Gulf War, the ultra-fine desert sand had done a number on the Apache power plant, sucked in through the filters and scouring the engines. We got an emergency call: could someone come over and take a look? As it turned out, the soldiers had already figured a solution: women's panty-hose had just the right mesh to trap the sand, and the soldiers were beseeching wives and girlfriends to come to the rescue. Thus it was, for a short time, until new filters could be provided, that the McDonnell Douglas Helicopter Company went into the panty-hose business, buying them by the carload and flying them into Kuwait.

You might ask, of curiosity, why are all of the big shows overseas? Because that's pretty much the way our aerospace industry wants it to be. There have been efforts by indigenous promoters, several in fact, to set up an "American" air show, but no one wanted to play. The U.S. market was already in hand, and foreign customers didn't want to make the trip to, oh, Dayton Ohio, when there were multiple venues closer to home. As noted above, it would be nice to see fewer major shows, but everyone knows that, overall, the system will never change. There is too strong an element of national pride in hosting a show (the French have been at it since 1909), and the show organizers simply make too much money.

Competition

From my Golden Gloves disaster, I learned that you must know the rules of the game before playing. Then, when I was in the sales and marketing game, I learned that you also must be prepared for fights that have few "rules," as happens in some large international aerospace competitions. I learned this lesson early from an event two years before I joined the company. Forewarned became forearmed.

Let me set the stage: a major competition involving two brand-new fighter models. At stake was a sale of perhaps eighty aircraft for at least 2 billion dollars. Both fighters were well-suited to their intended missions but had little in common. Fighter A was a powerful but nimble air-to-air combat machine; Fighter B was powerful, but more like a truck, designed to carry an integrated radar/weapons system with a two-hundred-mile reach that would take out an opponent before he was close enough for a dogfight. A strong U.S. ally wanted to upgrade his Air Force. President Richard M. Nixon offered a choice, Fighter A or Fighter B. There were briefings, presentations, but no clear winner.

A fly-off was arranged at Andrews Air Force Base, just outside Washington. There was an agreed scenario, each plane to fly identical maneuvers within a twelve-minute envelope. A high speed pass, a low speed pass, high-G low altitude turns, Immelmans and other typical fighter-aircraft evolutions. Each plane would start with an equal load of fuel, eight thousand pounds. The scenario clearly favored Fighter A, just the sort of flying for which it had specifically been designed. Also, Fighter A had a thrust-to-weight ratio of one-to-one while the heavier Fighter B came in at point-seven-to-one. The team selling B knew that they were at a disadvantage, but they knew how to play the angles.

Fighter A went first; a confident pilot flew a flawless, almost leisurely program. Several times he went off in the distance to get lined up for the next maneuver. Meanwhile, Fighter B was parked off to the side of the runway, engines running at a high throttle setting. The pilot was burning fuel to get weight down to twenty five hundred pounds, a point where his thrust-to-weight ratio would about match that of A. As you will appreciate, this was not part of the scenario but was within the rule to "start with eight thousand pounds." The rule didn't say, "take-off with eight thousand pounds."

When it came his turn, Fighter B's pilot also flew a flawless program but never left the boundaries of the airfield and flew most of the maneuvers at a low altitude right in front of the grandstand, thus enhancing drama for

Senator John L. McClellan (D-Ark.), demanding, cowboy tough, resolute, honest, fair, and wise. My mentor and guide from 1966–1975. (Courtesy McClellan Collection, Ouachita Baptist University)

James S. McDonnell—"Mr. Mac" to all—founder of McDonnell Aircraft and a true aerospace pioneer, was chairman of the board of McDonnell Douglas Corporation when he brought me into the world of aerospace marketing, 1975. (McDonnell Douglas archives)

Working with the Senate Appropriations and Government Operations Committees included some atypical beyond-the-Beltway assignments—here, on the ground in Saigon, 1974. (Personal collection)

Chinese media interviewing John McDonnell, 1995. (Personal collection)

Nephew, Sanford "Sandy" McDonnell, succeeded Mr. Mac as chairman in 1980. (McDonnell Douglas archives)

Mr. Mac's son Jim, VP for International Sales, with whom I shared an office in my first corporate assignment. (McDonnell Douglas archives)

The day I became VP of Marketing in 1986, surrounded by the top salesmen. From left: Jim Caldwell, Jim Kelley, Charles Cassmeyer, Dan Gilbert, and Alex Marshall. (McDonnell Douglas archives)

Harpoon, sea- and air-launched missile developed in the 1970s, a mainstay yet today. "Saving" Harpoon from the budget axe was my first triumph at McDonnell Douglas. (U.S. Navy)

F-15—first flight was in 1972, first delivery 1974. (U.S. Air Force)

Harrier AV-8B, vertical/short takeoff and landing (V/STOL) mutli-role aircraft which grew out of an earlier British model. (U.S. Marine Corps)

F/A-18 pushing through the sound barrier; the shock wave—which starts at the nose and moves aft—made visible in high humidity. (U.S. Navy)

F/A-18 (dual-mission fight/attack aircraft, generally marketed internationally as a fighter-only, designated F-18). (U.S. Navy; PH3 Kristopher Wilson)

AH-64D Apache Longbow, the world's most capable multi-role combat helicopter. (U.S. Army)

C-17 Globemaster III airlifter, controversial gestation, happy birth; first delivery to the U.S. Air Force was in 1995. (Department of Defense; SSgt. Jacob N. Bailey, USAF)

KC-10 Extender, an aerial tanker based on the DC-10 passenger jet. KC-10 entered service in 1981. (U.S. Air Force photo by Staff Sgt. Jerry Morrison)

MD-11, successor to the DC-10. First delivery 1990, last 2001. (Boeing)

JDAM: Joint Direct Attack Munition, attaches an intertial or GPS guidance kit to a range of "dumb bombs" to provide exceptional accuracy. (U.S. Air Force)

Conferring with Alexander Haig, secretary of state emeritus, en route to Beijing for the 2nd China Aviation Summit, 1995. McDonnell Douglas had three aircraft assembly plants in China. I was then corporate VP for Business Development and International Relations; Al Haig was one of our consultants. (Personal collection)

John F. McDonnell, Jim's brother and Sandy's successor as CEO and chairman, 1988 until 1997. (McDonnell Douglas archives)

Back to Vietnam after twenty-three years, 1997, to explore cooperative opportunities. In Hanoi, from left: Tom Do (a manager at Douglas Aircraft), John McDonnell, me (in my role as president, McDonnell Douglas International), Nguyen Hong Nhi (chairman of Vietnam Aviation), Vo Van Kiet (prime minister), Mr. Trung (secretary to the prime minister), Mr. Do (interpreter). (Personal collection)

the spectators. He added a couple of moves not on the program, including a show-closing inverted pass at less than five hundred feet, followed by a quick flip to upright attitude, a tight reverse course and drop to landing, with a rollout less than half that required by A.

It was a spectacular performance, and when it had ended, the client—the Shah of Iran—never stopped to visit Fighter A, the McDonnell Douglas F-15, but headed immediately for Fighter B, the Grumman F-14, to congratulate the pilot, sit in the cockpit, and demonstrate that he had made his choice.

The F-15 team was, to say the least, astonished. They tried to file a protest with the FAA ("conducting low-level acrobatics over a crowd") but got nowhere; the F-14 team had cleared their program with the FAA in advance.

The Grumman guys knew the strengths of their competition (formidable) and knew of at least one weakness—overconfidence. They knew their own strengths (and weaknesses) and how to work with and around them. I'd say pretty good starting point for any competition. In the event, the Shah got an airplane that served his needs, focused on keeping Soviet MIGs at bay. Seventy-nine of the airplanes were delivered with number eighty ready to go when the 1979 Iranian Revolution forced the Shah into exile. At that point, the United States stopped providing support, and without a steady supply of spares and technical assistance the Iranian F-14 fleet has dwindled to insignificance.

But the lesson learned remains: take advantage of any advantage. For example, we volunteered to take on the chore of writing a periodic report for an organization of aerospace industry contractors, which was distributed to all members and passed along the current and future plans of various key DoD offices. The modus was simple: the aerospace organization contacted the offices, requesting time for an interview with their editor—our man—who had the reportorial skills of a vacuum cleaner; he scarfed up everything. He would send his draft copy to all of the interviewees, accept corrections, and publish. This was all very open and above-board, everything our man learned (except for what was removed as "corrections") went into the report. This cost us a bundle—the writer's salary alone was in the six figures—but the benefit was priceless. Yes, he may have picked up some intelligence that was not included in the report, but the real value was that it gave our man face time with all of the most important officials in DoD, under cover of working for an independent group. As a result, he was often the guy they would go to for information or guidance on industry activities. Underhanded? Anyone could have volunteered for the assignment but no one else did. You be the judge.

Another "fair advantage." When the personal computer (PC) first came out we loaded a few with a modeling program we created that showed how our airplanes stacked up against the threat in any given country, gave the computers to potential customers in those countries, and sent along a couple of technicians to demonstrate the system.

Or, suppose there is a competitor's program that's causing you fits. You wish you could get someone in Congress to slow, if not derail, the whole useless thing without leaving your fingerprints. Well, don't enlist the aid of your hometown Congressional representative, but go to, say, North Dakota, armed with a bunch of campaign contributions from some of your North Dakota suppliers (there will always be some) and ask a local member for help in locating a potential site for a new labor-intensive overhaul and repair facility.

Or you hear that a member wants to cut back on the buy of your most important product, "The nation can't afford that many." Prepare some ammunition and get some of *your* friends in Congress to start a firefight, "This airplane is too vital to the nation's defense, but we indeed can save as much, or more, by cutting back on the acquisition of obsolete fighters." Fighters built, of course, by your competition.

Yes, I jest. Sort of.

Think "gamesmanship." You want your opponent to think you're going to do one thing, and you do another. Like the sport of fencing with a feint, parry, disengage, lunge. I've used straw men, for example, running an ad extolling our strength in something, which forced the competition to focus on what may actually be a minor feature. He had to divert resources from pushing his own strengths. I've also held a press conference for the trade press, so that I could be quoted: "McDonnell Douglas is sick and tired of offset programs where we have to give all this stuff away and don't get anything back." Your opponent puts his guard down, thinks you're not going to play this game when you really have offered the most sophisticated giving package anyone could ever want. When the truth comes out—after you win—his only excuse is, "But, we heard that Gunn himself said. . . ."

Some international competitors have had great success in getting senior political leadership to step up and support the cause. France and the U.K. have been particularly adept at this, and even the "Iron Maiden," British PM Margaret Thatcher, once flew to Saudi Arabia to make a direct, personal pitch to the king (page 131). Getting the U.S. government to provide similar assistance was sometimes iffy, because good old American politics often got in the way. However, with balance of payments issues and all, the U.S. government soon enough came to favor massive exports, the more massive the better. Embassies now have commercial desk officers to provide advice and assistance and U.S. government officials step in to remind potential customers of the constraints of the Foreign Corrupt Practices Act (FCPA): any violation and our government would withhold permission for release of technology. This helped level the playing field especially when the viable competition was only between American companies. In essence, "If you don't play by the rules, you get nothing." That was okay with us, even if no one got the sale, because the alternative was to let a competitor buy the contract. It's been known to happen.

Anytime we were in, or about to enter, an international competition, we made sure that our government leaders knew everything they needed to know, especially those who would be meeting with the customer's officials. Here are the issues, here are the solutions, and here are the sticky areas. From the feedback I got, this proved helpful whenever questions came up. For a major procurement, they usually did.

As noted, a few years before I came aboard, the company lost the Iranian competition to the Grumman F-14. A few years later, we got back at Grumman when we put the F-18 up against the F-14 in a Navy competition. And won. The word on the waterfront was that Grumman had provided such a long string of winning airplanes, World War II to post–Vietnam, that the company owned Navy air. They were all excellent airplanes, and the F-14 was an excellent airplane, but it cost a bundle to operate. We're not talking about fuel or upkeep but about one simple and often overlooked factor: it took two people to fly the F-14, the pilot and the naval flight officer (NFO) who handled navigation and weapons. The pilot was a given. Every manned airplane needed a pilot, including the F-18, but thanks to the miracle of modern technology, the F-18 pilot could do everything without someone sitting in a rear seat. As it was, the Navy had to train, maintain, care for, and provide retirement benefits for enough NFOs to man the fleet, be off in school, and offer a sufficient number of non-flying assignments to provide shore duty opportunities for everyone on some predictable schedule. That alone mandated at least two NFOs, maybe three, for each F-14 in the fleet. The Navy could get a new airplane with improved capabilities at only about half what an F-14 would cost over the life of the airplane. You get the idea. Think outside the box.

A similar argument was made for the McDonnell Douglas C-17 against the Lockheed C-5, with a crew of three (pilot, co-pilot, and loadmaster) versus a crew of six (pilot, co-pilot, two flight engineers, two loadmasters). I recall a 1985 Air Force study that estimated that the C-17 would require fifteen thousand fewer personnel than an equal number of C-5S.

I studied the competition as if I was cramming for the bar exam and collected and kept all sorts of information about our competitors, gleaned from myriad sources, public and private. We talked with pilots who flew the airplanes (ours and those of the other guys) to get their thoughts, good and bad, especially bad. We kept track of things on the rumor circuit. We collected press clippings and brochures. We may not have uncovered anything earth shattering, but the cumulative effect was powerful. I never used any of this information until it showed up in the media, usually the trade press, as it always did.

We worked up very detailed comparison sheets, showing the strengths and weaknesses of everything in the marketplace, including our own. (We, ah, didn't let the customer see our weaknesses.) Some comparisons were generic: the Russians could boast a lot of power, but in our judgment it was "air show" power, great for flying a pattern in front of the crowd but not always useful in tight combat. Most comparisons were specific: one of the European offerings had great performance and was cheap to buy and operate but was old technology; another had great performance and was as modern as they get but was very expensive and had a spotty sales record, which meant that downstream support might become difficult.

Our goal, of course, was to uncover advantages over the competition. It didn't always work out that way, and you may imagine my surprise when I discovered that the hated F-16 had been fitted out to carry our Harpoon missile! Harpoon capability was one of the strong features we pushed when selling the F-18 and it always added a few points in the evaluator's scoring. The logical question: how could it be that the F-16 was now in the Harpoon club? The astonishing answer, because another division of McDonnell Douglas was hired to develop the appropriate launcher and interface for General Dynamics, and nobody thought to tell the aircraft division! Call it a left-hand right-hand sort of thing; the bigger the company, the harder it is to keep anyone informed of what everyone else is doing. Had I learned of this earlier, I'm not sure I could have or should have done anything to stop it, but it would have saved a lot of heartburn.

The British and the French are normally our partners in international affairs, but when it comes to business partnerships, they have some pretty foreign concepts. How do you split work share with a company that has, for example, a nationally-mandated 36 hour work week with no overtime, where moving workers is almost impossible, and relocating facilities is definitely off the table? The correct answer is, very carefully.

Another issue: while businesses in Europe may or may not be nationalized, government subsidies are common and complicate competitions by allowing a manufacturer to offer a lower than realistic price. On a partnership deal, how do you factor their subsidies in with your work share; who is investing how much and for what? But I must admit American companies carry a bit of difficult to overcome baggage: the export license. More than once when working on a possible partnership I discovered that the U.S. government had no intention of allowing us to share some important bit of technology. Here I was, asking someone to invest in the project but having to mention, oh, by the way, you won't have access to the key information. The U.S. government's position, articulated by DoD and state, is that we must protect the crown jewels of our national defense. But in today's world, we are partnering for area defense, not just this nation or that, but for defense from a common enemy. When I've questioned a decision, I get something about "those people have a lax attitude about secrets." That may be true for some people, and caution is merited, but in my experience, 25 years of dealing with the Europeans, never once did I find anyone violating technical agreements.

Closer to home, natural and healthy rivalries between, say, the Air Force and the Navy are aggravated by sniping from their supporters (and your detractors) in the Congress. And perhaps, after his last big defeat, your competitor did not just crawl off to his cave to lick his wounds. Perhaps he can't upset this deal, but there will be others in the future. As an industry insider, he will know soon enough what issues arise, and know them before your client. Meanwhile, back at your home office, program management has been transferred away from business development and it is hard to know who now is responsible to do what, to whom, and to what purpose? And . . . what can you, or anyone, do when the customer is the U.S. Department of Defense and is about to take away the contract on an admittedly-troubled program and give it to your major competitor?

You go to general quarters and go as I did to a deputy assistant secretary of defense for acquisition and technology who was about to cancel our contract for the C-17 and give it to Lockheed. You propose a wrestling match, two falls out of three. Well, call that a figure of speech, I didn't actually throw him to the floor. I almost got fired, but I won the argument (full details, as part of the C-17 story are on page 113).

Espionage

It is inevitable; when you're dealing with aerospace sales in the billions of dollars and hold the key to some nation's national security, that you will someday or on many days be targeted by an industrial competitor or the spy corps of a nation, friendly or otherwise. They will be seeking an advantage, a technological hint, source codes, pending contracts, pricing, and your secret vices. Not only in Paris or Beirut, but as close to home as any city in which you have offices or operations, whether in Iowa or Indiana. Industrial and other espionage is an equal-opportunity threat.

Don't leave company-private or company-sensitive materials in your hotel room. They will most likely be compromised. Almost every one on my team reported evidence that someone had been rifling through their things, or had gone through their luggage even before they got to the hotel. Your company security department will have good advice on protecting data stored on a laptop (triple-encryption may be okay; not storing data on the laptop is better—unless, of course, you *want* someone to read your apparent secrets, perhaps prepared for that specific purpose).

Assume that you are always under surveillance of one sort or another. In some countries there's nothing hidden about it at all; you will be met at the foot of the boarding stairs by a phalanx of security guards who will take turns tailing you for the rest of your trip. If challenged—politely, of course—they will say, it's for your protection from bad elements.

I was once asked by a high-ranking customer in the Middle East to view some videotapes that showed some members of my staff meeting with persons the customer—the government—didn't want us to see. I got the message. Tell my people to break off the contact, at least while in-country. On another occasion, a national security agency played me a tape recording of a conversation I had with a Middle Eastern Sheikh over dinner at the Playboy Club in London. The tape had been discovered during a special intelligence operation in Syria, three years after the event. The people who played it for me assumed it was the Sheikh who was being monitored and wanted verification that it was me and this particular member of a royal family on the tape. They told me it had been recorded by a long range microphone probably concealed in a fountain pen.

Watch out for overly-chummy strangers such as the young man who pats you on the arm, friendly-like, and asks, "Hey, what city you from? My cousin lives in Chicago." He's probably a pickpocket. Maybe he wants your wallet, or your hotel key, which you probably still have in the paper sleeve on which the desk clerk wrote the room number. Perhaps that sultry brunette really does just want directions to the nearest Metro stop, but it's quite likely she wants to strike up a conversation . . . and who knows where that might lead?

Speaking of compromising situations—you will find temptations galore, many having nothing to do with sex, especially if you are posted overseas. A friendly local businessman, who knows you don't have a big salary, will offer to pay for a trip so your wife can visit your daughter in college. Another good friend will offer to bring you in on a hush-hush investment opportunity, or, how about a great job offer? Someone with exactly your expertise and background is needed (exactly that: they hope to tease out whatever you know about your company and products). These offers may be well-meant and totally legitimate, but perhaps not. All of these scenarios happened to members of my team.

At a dinner function in Moscow, I was given a "token of friendship," a wristwatch. I had it checked out by an intelligence specialist, and, what do you know, there was a tracking chip inside. Just like the old spy movies, where someone with a "reader" could follow me anywhere, and find out who I was seeing. About a year later I was in Tel Aviv having lunch with the president of one of the leading Israeli aerospace companies, and I noticed he was wearing one of the same watches. When I told him my little story, he almost fainted. He had been given the watch as a gift and had been wearing it for more than a year.

Such primitive devices have long since been overtaken by technology. Your cell phone probably sends off GPS tracking data—and the cell phone calls can be monitored. If you don't want anyone to know where you are going or what you are doing, leave you cell phone home. Or at least at the hotel

Don't use the room phone either. I always considered that rooms we used could be bugged and had our field offices swept every month. Probably not all that effective, but it made us feel as if we were trying. Whenever I checked into a hotel, I would ask to see the room I was assigned and then mumble some excuse and have it changed to another room. On one occasion, preparing for a private meeting with Serge Dassault, in Paris (page 129), and thinking I was defusing a listening device, I blew the major circuit-breaker for the wing of the hotel in which I was staying. This triggered the loudest darn alarm I've ever heard. I quietly packed my bag, checked out of the hotel, told the desk clerk I had a sudden change of plans. I didn't mention the noise and scheduled the meeting for another venue.

Another time, we learned that every room in the newest and finest hotel in one capital city had been wired during construction. One of our salesmen took advantage of that knowledge. He would go up to the executive lounge, get on the phone, and essentially negotiate with the customer via the hidden microphone while talking with someone in St. Louis: "I can meet their delivery schedule, but we're stuck on price, I've told them the best I can do and you know we can't go any lower, that's just a hard fact, and I can't hang around here much longer. I told them I have to have an answer by day after tomorrow, or I'm headed home and will have to release production to the next guy in line. If they change their minds too late, the best I could do was offer delivery two years farther out." He made the sale.

When I visited that city, I always stayed at an older hotel, built before the government got clever. There were—possibly—no bugs in the rooms. The government spies would have to listen in with less-than-ideal long-range electronic listening devices. I always had the radio or TV on, and water running in the bathroom just like in the movies. I don't know if it did much good, but it made me feel safer.

Overkill? Was I living in fantasy land? In 1993, the CIA obtained a list of fifteen U.S. aerospace companies that had been targeted for espionage by the French government; we were among them. That list focused on space and communications systems, where the French were trying to play catch-up. Here's an even more salutary lesson. In 1979, the State Department announced plans to build a new embassy in Moscow, and, as a gesture of goodwill, said they would use Soviet materials and Soviet construction workers throughout. The Soviets turned the embassy into a giant espionage collection device. They planted bugs inside its walls, used its structural steel for antennas, and even turned a nearby church into a listening post for transmissions, nicknaming it "Our Lady of Perpetual Observation." It wasn't until 1984, with the embassy half-finished, that United States counterintelligence caught on and plans were changed.

If the U.S. government can be fooled, well, so can you.

Beware, be ethical, be safe, and be happy.

RFP

The U.S. Army issued the RFP for its first airplane in 1907. It was pretty straightforward: the "heavier than air flying machine" should be easily assembled and disassembled (to be carried on an Army wagon), must carry two people weighing a total of 350 pounds at a minimum speed of forty miles per hour, be able to fly 125 miles, land safely on a field even if the engine failed, and should be "sufficiently simple in its construction and operation to permit an intelligent man to learn to fly the machine within a reasonable length of time." A flight demonstration was mandatory. This was a maximum-fee contract, 25,000 dollars, with positive (and negative) incentives. For each mile between forty-one and forty-four mph, there was a 10 percent bonus. For each mile under forty, a 10 percent deduction but if top speed was below thirty-six mph, the government would reject the entry. To limit participation to serious bidders only, each had to make a deposit equal to 10 percent of their asking price, to be forfeit if their airplane failed to meet the specifications.

There were forty-one respondents; most were deemed to be kooks. Three bidders sent checks. One of those soon dropped out, one was a fraud (his bid was 20,000 dollars, but he expected to award a sub-contract to the Wright brothers to build his entry; they were not interested). The Wrights won. They were not surprised; the RFP had literally specified the airplane they planned to offer. It did so because they had tried to sell it to the Army earlier, but the Army decided that to be fair to other airplane manufacturers (as if there were more than a couple) they should put it out for bid.

The two-page Army RFP is often cited by writers, reaching for humor, as a snide put-down of what they view as over-stuffed, over-specified, densely-written modern RFPs. They miss the point. It should be honored as an ancestor, not a giant-killer. It sets forth all of the basic elements of a good RFP. Short and to the point? Indeed, but that doesn't impeach any RFP of today. The world—and airplanes—have changed a lot since 1907.

Fast forward to a recent RFP, calling for a "full and open, best value" competition. ("Best value" simple definition: a combination of price, performance, risk, and a lot of other things; it offers the government greater discretion in source selection than the contrasting "lowest cost, technically acceptable" method.)

What's involved? "The contractor shall accomplish all the effort necessary to design, develop, test, and certify the New-X military aircraft." The RFP sets forth the mission, performance specifications ("Takeoff distance—ground roll, 35-foot obstacle clearance, and 50-foot obstacle clearance; ability to operate from a 7000 foot runway and gross weight—for maximum fuel load, limited to maximum takeoff gross weight, standard day and standard day +20 degrees C"). The RFP describes the project, the development timeframe, initial number of airplanes to be purchased, and the initial rate of production.

What is to be included? The contractor will provide spare engines, accessories, expendables, initial training (along with information for training systems, simulator development, and depot level maintenance), support equipment, technical manuals, license rights for software, and on-site office space for up to, say, fifty government personnel or representatives. Other details cover inspection and acceptance, deliveries, post-delivery performance, identification of the contract administrator; how to deal with changes whether submitted by the government or on the contractor's initiative; trip wires that trigger adjustments to price; and what to do if the money runs out (stop work, more or less).

You must include information on your own past performance, which becomes part of the formal evaluation and may be rated in one of seven categories: High ("Based on the offeror's performance record, the government has high confidence the offeror will successfully perform the required effort"); Significant, Satisfactory, Unknown, Little, and No Confidence. "Satisfactory" is not the kiss of death, just a stern warning to stay on top of things. "Unknown" and below, you're in trouble. You (and your suppliers) can brag on past success in similar programs but must disclose problems or issues; if you don't, trust me, someone will find out. Describe the problem, explain your actions to correct and demonstrate your success, and provide assurance that your have controls in place to prevent a recurrence in later programs.

Be aware of the power of the "informal" evaluation of past performance, based on the customer's own experience with your company, your company's reputation, and, not inconsequential, your competitor's efforts to discredit you. One reason for staying in constant touch with the customer is to refute rumors and correct misperceptions.

The RFP will specify one or another (or a combination) of funding schemes: cost plus incentive, cost plus award fee, firm fixed price, fixed price incentive firm, and so on. There are targets, as a percent of total contracting price, for participation by small businesses, veteran-owned small businesses, service-disabled veteran-owned small businesses, Historically Underutilized Business Zone (HUBZone) small businesses, small disadvantaged businesses, and women-owned small businesses.

These are just a few of the things that your team must keep track of and address when analyzing the request and drafting the proposal. And you wondered why you needed a large and broadly-skilled cast of characters. Everyone gets into this act.

As for the proposal, itself—some nuts-and-bolts marketing—pay close attention to many things. Keep track of how many times something is mentioned in the RFP. If it shows up just once in the section on Product Support, it may be of interest only to one guy off in a corner who works on support issues. When you find it also mentioned in Key System Requirements, Software Integration, and Pricing, you may assume it is of some importance to everyone.

And what is called for in the RFP may not always reflect what some of the contributors or evaluators want to see. One of your team may tell you, "My contact says they are looking for something that will go beyond visual range." An important capability, but when the RFP arrives, it's not even mentioned. Should you ignore it, or is this something you could include with your offer? There well may be one important person on the customer side who will respond favorably to this "extra." I've heard at least one evaluator say, when asked about a certain point, "You do whatever the RFP asks you to do. But here's what I want to see. . . ." The person who wrote that section of the RFP was looking for a quick read in an effort to streamline the review; the evaluator wanted a great deal of detail. You can do both.

You may not submit a free-form offering, packed with glorious language and stunning illustrations. To wit:

> The proposal shall be clear, concise, and shall include sufficient detail for effective evaluation and for substantiating the validity of stated claims. The proposal should not simply rephrase or restate the Government's requirements but rather shall provide convincing rationale to address how the offeror intends to meet these requirements. Offerors shall assume that the Government has no prior knowledge of their facilities and experience.

The Government will base its evaluation on only the information presented in the offeror's proposal. Elaborate brochures or documentation, binding, detailed art work, or other embellishments are unnecessary and are not desired. Similarly, for oral presentations, elaborate productions are unnecessary and not desired.

In days past, the RFP would specify everything from speed, ceiling, range, and crew size down to the size, shape, and color of the push buttons on the radio. Thus, competitors were bidding on almost-clones of the same airplane. In recent years, the rules have changed. The RFP may describe the mission and the contractors are allowed some latitude, with room for creativity, to develop an airplane that meets the requirements. This broadens the playing field, bringing in a variety of proposed solutions. Some folks miss the cookie-cutter approach, where the customer already has sorted out the specifications. It is a lot easier, relatively speaking, and takes a lot less manpower to come up with a competitive proposal. Of course, this is a well-worn path on the way to mediocrity.

The RFP spells out exactly what must be included and in what order; leave something out, you will be considered "non-responsive." (If the customer asks you to address a certain point in Section Six but you know it really belongs in Section Four, don't assert your independence and put it in Section Four. Put it in both.) You must follow the specifications for page size and format, type font and size, with all set to standard default values. No squeezing in a few hundred words here or there by varying the line height or letter spacing. Why not? Call it the level playing field factor. You will be limited to the number of pages in each section, say, twenty for the Executive Summary, seven hundred pages to describe Mission Capability (covering Key System Requirements, System Integration and Software, Product Support, and Program Management), fifty pages to brag on your past performance, and three hundred pages to cover all cost and pricing issues. Table of contents, glossary, list of acronyms, index are not counted in any total. Your incentive to conform? "If exceeded, the excess pages will not be read or considered in the evaluation of the proposal." That means, if your boss insists on cramming something in at the last minute, be sure to put the important stuff up front.

You may be permitted to make a "non-elaborate" oral presentation. Perhaps no more than four people may talk for no more than four hours, and be limited to one hundred slides. Do you still wonder why you need to have a dedicated proposal center, staffed with hordes of skilled writers?

On Staying Ahead of the Game

Let's say you're the head of Business Development. Here's a conversation you should never have:

"Boss, they just released the RFP on the Advanced Fighter Program."
"Great! What's in it?"
"I don't know, I just got it, haven't had a chance to read it yet."
Long pause, then you say, "You. Don't know. What's. In it?"

The customer has been working on that RFP for, oh, maybe two years. "Advanced Fighter" is not a hidden, black, top secret program, and your team has known about it for a long time. The person you have assigned to follow this should not only know who the players are and where their office is located, but *know* them. Know where they went to school, how many kids they have. Know them well enough to be able to drop by frequently for friendly chats or exploratory conversations. They may have ideas they want to explore where you can be of assistance. ("Yes, Mr. Icarus, your wing maker could save a bunch of money using wax, but for your planned operating environment he might want to look at epoxy.")

The RFP process starts by defining a set of requirements. Your job, your company's reason for being, is to help the customer fill requirements with appropriate products or services. Perhaps you have plans for an airplane that will fly at Mach 2, while you know a competitor's offering will barely top Mach 1.5. Since fighter pilots never have enough speed, it would not be out of line for your "Advanced Fighter" person—perhaps in this case he is a well-known, iconic, retired fighter pilot—to emphasize the importance of speed. Wouldn't it be nice if the RFP specified an airplane with a top speed "at least Mach 2"? You might call it the "Wright" factor: you have a head start when the RFP describes your specific product.

I'm over-simplifying a complex process, but you get the idea. In the actual world of military procurement, there are many players and you should work with them all. Start with the users, the war fighters, the folks in uniform who will depend on the equipment you provide to carry out a mission and stay safe while doing so. They will have their own wish-list—they want everything faster, brighter, lighter—which they pass on to the folks who actually set the requirements, who turn their package over to the acquisition guys who set the ground rules, evaluate the proposals, and will eventually buy the product. And don't overlook the maintainers who will keep it running once it is in service; what features are important to them?

These are separate and distinct communities that often do not see eye-to-eye. There may be cost-benefit, technology, timing, and post-delivery support issues to resolve at each step along the way. These present additional opportunities for the expert advice which you are free to offer, right up to the point where the RFP is published. And along the way, you might gently probe for unannounced customer needs, which will not be included in the RFP but which may play an important role in source-selection.

Here's a semi-hypothetical example (the basic story is true but simplified in the telling). You learn that one of the services wants to install new wings on an old warhorse—an airplane built by another company. The requirement will be a basic, traditional "form, fit, and function" replacement; the original wings worked just fine, they're just wearing out.

This will be a new customer. As a first step you need to introduce yourself to the folks working on or contributing to development of the RFP. They've heard of your company, of course, but might be rightly skeptical that you know anything about building wings for someone else's airplane and why would a big company like yours be interested? Set up a meeting, several meetings; take your own wing designer along to talk technical. Outline your experience with modifications and upgrades. Let them know you're really interested in this project; ask for any information they can pass on at this time. Mention how you recently incorporated some new materials into the wings of one of your airplanes, composite fasteners or whatever will last many years longer than the original, and, oh by the way, saved a lot of weight, leading to decreased fuel consumption and increased range. As a result, perhaps the narrow request for a "form, fit, and function" copy of the original will be modified to allow bidders to suggest improvements, if at no additional cost. You will have established your company as a qualified, motivated contender. I call that a pretty good start.

What about the un-formalized customer needs? Heavy maintenance and re-work are usually handled by a depot. It doesn't take too long to learn that there hasn't been much work lately for the depot, sure would be nice if they could somehow be involved. Why not put in your bid, that each set of new wings will be shipped to the depot in three kit boxes, for assembly and installation on the airframes?

Another win for the home team! More cars in the parking lot!

Now, let me move up a notch to a fully real-world example. By 1989, with the C-17 program underway, it was time to go on the hunt for foreign customers. A prime candidate was the U.K., but first we had to help the British convince themselves that, (a) they had a valid, long-range heavy-lift requirement; (b) it could not realistically be met by the then-planned European Future Large Aircraft (FLA); and (c) while loyalty to FLA-team member British Aerospace (BAe) and its European partners was commendable, it was also an encumbrance.

At the time, British heavy airlift was being handled on an ad hoc basis by leased Russian Antonev aircraft with a penalty in both cost (high) and prestige (compromised). Those factors were easily quantified, and raising the Russian connection helped soften (but not eliminate) reluctance to bypass BAe and European allies to buy American.

Our job became even easier in the mid-1990s, when the British created the Joint Rapid Reaction Force (JRRF). They already had elements for the "Force" but were missing the "Rapid" part. Further, the FLA (intended as a replacement for aging C-130s serving with eight European nations) would have a payload only about one-third that of the C-17 and simply would not be able to handle out-sized or heavy cargo. Also, perhaps as much to the point, by this time we were delivering C-17s to the U.S. Air Force, and the FLA was a paper airplane, with first-flight yet many years into the future. I do believe that the concept "rapid" includes adding the capabilities sooner, rather than much, much later.

When Tony Blair took over as Prime Minister in May 1997, he asked for a thorough review of British armed forces and missions. We made certain that the appropriate offices and officials had the pertinent data. In July 1998, the MoD issued a "Strategic Defense Review" (SDR), which included:

Item 96: **Strategic Transport.** It is not enough to modernize the front line. We have an urgent need to improve our strategic transport, to allow us to move more powerful forces quickly to an overseas theatre. In the short term we expect to do this through acquiring four additional roll-on roll-off container ships and four large C-17 aircraft or their equivalent. In the longer term, we will also need to consider a replacement for our remaining elderly transport aircraft. The European Future Large Aircraft is a contender for this requirement.

Note well: "C-17 aircraft or their equivalent." There was no "equivalent." It was about as close to a commitment as any government would likely put in an official public report, although I believe there was a lot of internal haggling before the MoD made a purchase decision. Parsing the word "equivalent" is, after all, a challenge.

For our part, to facilitate the transition, we offered a lease-purchase agreement for the four aircraft, the first such deal we had made, I believe, for a military aircraft sale, foreign or domestic. Ten months after publication of the SDR we had a contract, and the first aircraft was delivered 366 days later. Pretty rapid. The British later ordered a fifth C-17 as an outright purchase and announced that the purchase option of the lease-purchase agreement would be exercised in 2008. (I might note that the FLA, a program managed by Airbus Military and now designated A-400M, has yet, as of the middle of 2008, to make first flight. Potential customers have enjoyed walking through a full-size plywood and plastic mock-up parked at European air shows.) I don't mean to make our success with the U.K. C-17 sale sound easy or quick. Our team in England spent almost ten years developing the relationships, framing the arguments, and answering the objections. You say, only five aircraft? Does about 1.5 billion dollars mean anything to you?

Of course, even with the best sales team with the best product and the best offer, you won't win them all. After a loss, someone (the VP for marketing or an executive who was not part of the team) should visit the customer—not to complain but to learn. Thank them for giving you the opportunity to bid, with a "Sorry it didn't work out for us, but we want to remain (or become) a key supplier. Is there anything you could share with us—not the competitive specifics, but what did we do wrong, or could we have done that might have made a difference?" The question is not as important in a DoD competition, because they always hold after-action briefings to spell out the discriminators, but a non-confrontational courtesy call never hurts anyone. You spent a lot of time building relationships with the customer, and you want to keep them polished bright. Let other losers act like, well, losers.

I gained an invaluable insight from one such "courtesy call." We had been working Israel, hard. They had been using our equipment for some time—F-4s, F-15s—I knew all of the leadership, but in 1978 we lost a campaign to sell some more F-15s. They bought F-16s. How could this be? The Israelis always bought the best, top of the line, and to my mind there was no comparison. Around the shop we used to say (only half in jest) no wonder that the USAF was encouraging the air forces of Europe to buy F-16s, they couldn't hurt anybody and therefore would not be likely to start World War III. (You have to inject a little humor into your work once in a while.)

Well, I made the trip to Israel, had an excellent meeting with some of my friends who were most cordial—I'd known some of these guys since my days working with the Senate. But, to be honest, I hadn't learned much to add to lessons-learned until the Air Force chief of staff took me into his office and closed the door. The issue, he said, was price. They had a limited budget and were adopting a high-low concept. At this point, they had all of the F-15 high-end, long-range air superiority fighters that they needed or could afford and were adding a somewhat limited, low-end capability. The F-16 was well-suited for close-in combat missions such as dropping bombs and strafing enemy troops, and indeed was a good replacement for aging F-4s—if that was all you needed. The lesson learned in 1978, the hard reality, was that we were not going to dominate airspace with our primo fighters; but that the F-16 would always find a market and we had to deal with that.

Offset

Offset—more properly called, "industrial participation"—is reciprocal trade in order to secure a sale; a certain percentage of the sale price must be offered back to the buyer in some form of trade or training. If this sounds like a ploy to get around the FCPA (especially since the concept appeared just about the time the FCPA was passed) it is not; it is open, legally competitive, and in most offset deals, the buyer and seller both benefit. For example: the seller agrees to place orders with a local manufacturing company to make parts or assemble components for the airplanes in the deal. The seller would have to buy those parts from someplace, why not from a local supplier? The local supplier has to supply the work at a competitive price, so the actual cost to the seller is marginal, perhaps to supply a bit of training or an assembly jig, but the offset contribution may be calculated on the value of the orders, or on whatever the buyer and seller agree. The seller gets to sell some airplanes.

Offset should not be confused with "counter-trade," which is just another way of helping to pay for the goods. In the 1960s, the cafeteria at Douglas served up a lot of Yugoslavian ham, taken in trade as part of a commercial sale. In a 1985 deal to sell fighters and trainers to the Saudi Air Force, the British absorbed a lot of Saudi oil, measured in the billions of dollars and delivered at the rate of one supertanker a day for a long, long time. (For more information, go to page 131)

By the early 1980s, offset was all the rage, a major discriminator, and McDonnell Douglas had some success in offset-balanced F-18 sales to Canada and Australia. Offset was a new concept on both sides—the customers knew they wanted offset but had not quite determined what it should be. We sent in teams to help them decide what they wanted and what they needed, and worked out arrangements for export assistance, training, some technology transfer, and direct aerospace work. However, it did take the company a couple of years to fully get the hang of it.

Turkey wanted help in developing an aerospace industry, we wanted to sell F-18s and thought we could work something out. Unfortunately, we were a late entrant into the competition and when the sales team went to talk turkey, they learned that the Turks already had a detailed development plan, a roadmap, a primer on "how to set up an aerospace company" and floor plans of the plant and offices, courtesy of General Dynamics. Needless to say, Turkey bought F-16s.

We thought we had worked out a good offer with the Greek government when, at the eleventh hour, they said they also wanted McDonnell

Douglas to make a 50 million dollar investment in Greek industry. We interpreted that as an under-the-table anti-FCPA transaction and turned it down. General Dynamics set up a venture capital fund, to be invested over time, with General Dynamics and General Electric on the board of directors. They won.

Spain had the world's largest olive harvest, but large quantities of Spanish olive oil were sold in bulk to Italy, where it was mixed with Italian olive oil and sent to the United States as "Italian." We proposed arranging for the direct import of Spanish olive oil, creating a U.S. market and eliminating the middlemen. Not a good idea. It turned out that the U.S. distribution of Italian olive oil was controlled by a Mafia family and they did not look forward to encroachment on *la sua cosa*. We changed direction. Another bright idea: we offered to help Spanish shoemakers establish an American market, until the St. Louis chapter of some international brotherhood of shoemakers and a local shoe company found out and went crazy.

Two lessons learned: First, we didn't belong in either the olive oil or shoe business. The deals looked good on paper but ignored reality, and, second, we shouldn't be doing things with offset that made it look as if we were desperate, willing to do anything to sell an airplane.

Our philosophy of offset evolved, and we moved away from schemes. Some, like hi-tech windmills or plants to process titanium sponge, may have seemed interesting, but had already been bounced off company after company, bank after bank, and had no rational chance for survival as a stand-alone business. Some folks thought, well, we don't care, McDonnell Douglas will give us enough support to make it work. Sure. Some folks didn't understand the underlying realities. We focused on programs that made business sense, especially aerospace-related activities, manufacturing, training, internships, and special projects.

Soon enough the company did indeed get the hang of it and sold seventy-two F-18s to Spain. In fact, I believe that Spain was the first country to have a full-time offset program manager, an offset czar, who entered into his job with gusto. Did our deal include training or interaction of one sort or another to be provided over a given period of time? He would squeeze every possible hour of training or consultation or coordination, and would not let us get ahead of schedule. The program ran for the entire length allotted. Plus the grace period. The man really knew how to do his job.

To this point, most of our offset-related business had been in the industrialized countries of Europe, Australia, Canada, and Israel, all of which had qualified suppliers. Things were quite different in less-developed nations. If we found a company with a good reputation but lacking the right technology or skills, we would transfer releasable technology and provide some training. This put that company in position to compete for our business, but they earned no premium: to win, they had to have a winning bid. For example, we helped a company in the Czech Republic learn how to make F-18 gun bay doors. They bid, they beat the competition, and they got the contract.

In the Middle East, everyone wanted offset but there was little indigenous industry. Most of the manual labor force was brought in from third countries, and there were not many—any?—qualified ready-at-hand suppliers. In Saudi Arabia, we set up partnerships with the government for aerospace maintenance, software development, and to produce miscellaneous aircraft equipment. We created training programs for the workers, which gave many a new purpose in life and helped combat endemic unemployment. Most of the training was conducted in the United States until the fall-out from 9/11 complicated entry for students from the Middle East. Training was then shifted offshore to Saudi universities, the Australian-Kuwait University and other facilities, which offered classes in a broad range of basic technical skills.

To identify in-country assets, where there were some, we held conferences with local business leaders and company managers. In a matter of days, we would collect data from perhaps two hundred companies—their size, capabilities, experience, areas for growth. The offset team would analyze and organize the information, then visit companies of interest and form alliances with some. McDonnell Douglas got friendly coverage in the media and high marks from the offset authority and politicians. Here was tangible evidence that this big American company was reaching out to provide jobs, training, and technology. A nice head start when it came time to make our pitch to the customer.

Some customers know exactly what offset they are seeking. The more sophisticated among them want access to the latest technology and co-production agreements, whereby after a break-in and training period, they will be manufacturing portions of the airplanes and assembling all of them. "Wait," you might ask, "doesn't that reduce the value of your sale?" It certainly does, by a significant amount, but we wouldn't have had the sale at all without the agreement, and our workers and facilities were freed up to take on other contracts.

Does co-production worry the unions? Sure, when we signed our first contracts the unions thought we were giving away jobs, but we explained that, by winning a competition because of the co-production agreement, we preserved or added jobs that would have gone to another company. We took that message to union management and to the union halls and I think most of the union members understood. Once in a while a union official would make a public statement, something like they wished they had 100 percent of the jobs instead of 90 percent. For their own reasons, union officials usually are careful not to appear too close to management.

Some customers don't want to help make our airplanes, they want us to help sell some of theirs, invariably, trainers. Their eyes are bigger than their stomachs; they convince themselves and their governments that they can handle simple, basic prop or jet trainers and just *know* that a market is standing ready starting at home. Of course, with so many companies vying for a discrete trainer market, something has to fall out. General Dynamics tried to help the Taiwanese; the only sales were in Taiwan. They worked with the Koreans. The only sales were in Korea. One of the most difficult parts of "offset" is to dampen the enthusiasm of the naive, and to tamp down improbable expectations.

Not all offset involves manufacturing. For the Netherlands, we set up an internship program through Delft University, whereby aerodynamicists in the Masters or Ph.D. programs would come over to work alongside our people on some real-world problems. In a ten year period, I think we hosted some fifty very bright, very technical students. Some made major contributions, and some were even granted U.S. patents.

Here's a true and continuing success story. Some of our advanced manufacturing people in St. Louis had established a relationship with a professor at the University of Sheffield, England, and we were looking for some offset possibilities. Sheffield was the birthplace of stainless steel and had been a center of coal mining, but had fallen on hard times. How about establishing an Advanced Manufacturing Research Center (AMRC), a partnership between the company and the university? The local and regional development agencies said yea, Parliament was interested, Prime Minister Tony Blair backed the idea, and we had a deal.

What began as a modest effort became the seed for a one-hundred-acre technology park (Prince Andrew participated in the grand opening) in which we are still involved. Machining techniques developed there allowed a major landing gear manufacturer to propose a lower weight, lower cost titanium landing gear for the Boeing 787. They won the contract and Boeing got a great landing gear. You might call it harvesting the benefit from a modest investment in an offset program from which the company already had great benefit.

Governments change, people change jobs, and people retire. Some customers, even today, might not have a clue about offset other than what they read about it in the trade press and that they want it. This takes us back to the beginning—the need to understand the customer's wants, needs, capabilities, and culture, all of which require on-the-ground time and effort and perhaps a consultant or two. In other words, a lot of the things you need to know in order to sell some airplanes in the first place. In the early years, the offset team was part of the marketing department but coordination within the department was spotty. It became apparent that the sales and offset teams were chasing the same data, the same contacts, and the same goals. Better communication led to better coordination and a more effective effort all around.

Don't forget an often overlooked key benefit of offset: some offset programs may run for eight, ten, or twelve years, and continue to run long after the last airplane has been delivered and post-delivery support has ended. Through the offset, you will still be involved, still working with the customer and suppliers, keeping relationships solid for the next sale.

CHAPTER 4

Selected Case Studies

This is not an official or comprehensive history of any transaction or program; rather, in the time-honored mode of "don't just tell me what to do, show me how to do it," this chapter puts a focus on selected work of the sales and marketing teams that I was involved with in one capacity or another. These case studies include examples of airplanes, weapons, and support services, presented more or less in chronological order—although there were many overlapping programs and many programs are not included here.

Here you have what we did, why we did it, how we did it, and to what result in a range of complex and, at times, apparently doomed efforts. The details are based in large part on our collective memories, which may differ at times from whatever is on the published record, wherever appearing. Some of the things that we did, or that happened to us, are most assuredly *not* on the record. They may be the most instructive parts of the narrative.

C-17

The most successful program in company history almost didn't happen, several times. Call it an example of how many things can go wrong and how often, before everything comes together. This tale begins in 1973 when Boeing and McDonnell Douglas each received a contract to build and test prototypes for the Advanced Medium STOL (Short-Takeoff and Landing) Transport (AMST). The idea was to develop a modern replacement for the aging C-130 cargo plane and test some new technology for short-field landings and takeoffs. Boeing entered the YC-14, and McDonnell Douglas entered the YC-15. After some successful flight tests, the program was cancelled.

I was then the director of government affairs. I found out about the cancellation from the director of the Office of Management and Budget (OMB), who called me and said, "I've got bad news and good news." The bad news: the AMST is dead. They were going to re-open the Lockheed C-5 line—the giant of airlifters—to build fifty more airplanes. The good news: as a consolation prize McDonnell Douglas would be awarded a contract for forty-four new KC-10 dual role tanker-airlifters.

"Budget reasons" were cited as the official excuse for the cancellation. The AMST would be too expensive. The AMST wouldn't resolve another airlift issue, namely, what to do about the much-larger but also aging C-141s. Both statements were true. However, I was told by a knowledgeable official that the administration didn't want to replace aging C-130s with anything but new C-130s, and the replacement for the C-141 would be an improved version of the C-5, the C-5B. Both airplanes were built by Lockheed, in Georgia, home state of President Jimmy Carter, who happened to be the direct boss of the "knowledgeable official." Perhaps it was just a coincidence, but neither of the AMST contractors had operations in Georgia. Or so I was told.

DoD and the Air Force agreed with the decision to cancel the AMST, but also agreed on an urgent need to study the airlifter issue. This led to the RFP for a new airplane, the C-X (Cargo Transport Aircraft-Experimental). C-X could cover the short-range (tactical, inter-theater) missions of C-130 and the long range (strategic, intra-theater) missions of the C-141, but do a better job, carrying more cargo, with a smaller crew, and unlike the C-141 and the C-5, operate into and out of small, austere airfields closer to the action. The C-X was a much-needed gap-filler.

Three companies entered the fray: Boeing, which proposed a modified 747 freighter, Lockheed which proposed the C-5B—both operating in the "let's get them to buy something we already have" mode—and McDonnell

Douglas, which proposed a totally new design based on, but much larger than, the YC-15 AMST.

A team of our marketers made the rounds on Capitol Hill, briefing key members of Congress. The competitors were doing the same thing, but I think we were a bit behind. When our team dropped by to talk with Senator Barry Goldwater, major general USAFR (retired), he asked to see a photo of the airplane. Our team didn't have a photo, just an artist's concept. Goldwater brushed them off. The other people had been by with photos. No photo? Well, then it's just another paper airplane.

However, toward the end of 1981, McDonnell Douglas was picked as the winner. First big error: the new (Reagan) administration wanted to control defense spending in a manner as to keep money available for more projects. DoD elected to issue a fixed-price contract rather than the cost-plus funding typically applied to new programs. DoD, looking at the YC-15, claimed that all of the technology had been proven; the fixed-price would hold the contractors accountable and reduce excessive overruns.

Well, maybe. The McDonnell Douglas YC-15 was a demonstrator, not an operational aircraft. It had a range of four hundred miles and a payload of 27,000 pounds, compared with a planned C-X payload of 172,200 pounds, to be carried more than 2,700 miles without refueling. This chicken would soon enough come home to roost.

But no sooner had the selection been made than the program was put on hold. DoD announced that geopolitical urgencies required new airlifters sooner than the C-X could be developed and delivered. The fifty C-5B / forty-four KC-10 alternative was made official in January 1982. Boeing rushed in with an (unsuccessful) unsolicited proposal, offering fifty 747s. There were rumors of collusion between Lockheed, DoD, and the Air Force, to prevent Congress from overturning the "alternative"; at least, such was the opinion of the General Accounting Office (Betty Raab Kennedy, "Historical Realities of the C-X Program Pose Challenge for Future Acquisitions," *Proposal Manager Magazine*, November–December 1999, pp. 72–80; a bimonthly magazine of the Defense Systems Management College).

Well, perhaps Congress was listening, because they denied FY82 development funding for the C-X. However, thanks to some fast footwork on our part, Virginia Congressman Dan Daniel agreed to slip in a line-item that preserved a couple of million dollars for the study of "airlifter requirements and alternatives to the C-X." This gave us enough of a toehold to hold on—work our contacts in DoD and the rest of the Congress—and keep the program alive.

We slowly built up a constituency. We planned to build the C-X at the Douglas facility in Long Beach, California, which had capacity to spare. We hoped to put as much of the work as possible into Long Beach, but political realities intruded when the chairman of the Senate Armed Services Committee told our CEO that he would have a problem supporting such a big program that did not have jobs in his home state. We realized that he was not alone and signed contracts with major suppliers in several states that had elected officials on key defense committees.

Some members of Congress found it hard to believe that an aircraft that did not shoot bullets or drop bombs could be decisive in a war, so we developed some "Win the War" scenarios using an Army-approved program. The most dramatic was in Korea: Seoul was lost without the C-X and saved with the timely delivery of rocket launchers (coincidently made by one of our suppliers, which had created the program that created the scenario; as I noted earlier, there's nothing wrong with looking for an advantage).

In September 1983, the Air Force showed renewed interest in the C-X. The chairman of the Armed Services Committee—perhaps now firmly in our corner—asked the Government Accountability Office (GAO) to study "airlift," specifically comparing the C-X with the C-5. In a report issued in July 1984, the GAO affirmed the need for new airlift capabilities and suggested that the C-5 was not a viable candidate. Lockheed claimed that the new C-5B could handle small, austere airfields just fine; the Air Force told the GAO, probably not. There would be no margin of safety and the airplane was too big for the limited ground operations provided at small, austere air fields. The DoD came over to our side and put the C-X, now officially the C-17, on the path to full-scale development. Lockheed, nothing if not tenacious, submitted another unsolicited proposal asking to complete austere airfield tests that had been suspended in 1970. They had been suspended for good and sufficient reason when some austere field-connected problems developed. The test was not re-opened. Full scale development was authorized in 1985.

In January 1986, a paper published by the Heritage Foundation charged the Pentagon with "making a serious and costly error" by pursuing the C-17 because there was sufficient airlift capacity already and most of it was underutilized. "Careful analysis by experts of U.S. airlift needs and of the C-17 program reveals that a new cargo plane is not needed to close the gap." Besides, the author wrote, "some experts see problems with a hybrid design that equips the C-17 for both strategic and tactical airlift missions." The author did not identify the "experts." He asked, "Is it realistic to expect the Air Force to risk the C-17, which may cost $180 million or more each, on austere airfields in or near combat zones?" His recommendation: cancel the C-17, build more C-5Bs and KC-10s, and keep all of the C-130s and C-141s. This would save $20 billion, he wrote. I'm not sure if the author ever took a close look at the condition of even an average C-141. And, oh by the way, he suggested that the Air Force might want to develop a new "tactical" airlifter.

I don't know where the author got his information or his bias (I could hazard a guess) but call this a great example of the media trying to influence policy, although the paper did carry the caveat, "Nothing written here is to be construed as necessarily reflecting the views of The Heritage Foundation or as an attempt to aid or hinder the passage of any bill before Congress." (Kim R. Holmes, "The Military Airlift Gap," *Heritage Foundation Backgrounder* 482, 23 January 1986.)

In 1987, a Congressman from Georgia (yes, home of the C-5) tried to delete all C-17 funding. That didn't happen, but over the history of the program Congress cut the funding and changed the number of airplanes several times. Each time, of course, raising the unit cost, which led to another round of cuts.

The first flight was in 1991. This was behind schedule. In June 1992, the secretary of the Air Force warned John McDonnell that the company had to get its act together. In October 1992, the wing failed a scheduled overstress test. Well, that's what the test was for, stress the wing to 150 percent of maximum expected load and see what happens. It failed at 128 percent. After modifications, another failure, and then after more modifications, the wing passed. In the interim, reading some of the press, you would have thought that the C-17 was going to be a flying disaster. An on-site GAO team was too free with semi-accurate information, and a local Long Beach reporter was practically camped out at the plant. If someone dropped a wrench it was likely to show up in the newspaper as a major industrial accident. (Yes, once again I exaggerate, but you get my drift).

This is why you need an experienced public affairs/PR staff, with direct access to the boss and the authority to do what should be done if the boss is out of range. Our man on the C-17 was a retired Air Force Public Affairs Officer (PAO) with previous commercial aerospace experience. He didn't wait for the story to leak out around the edges but called the media and arranged an on-site briefing explaining the test and outlining the solution to the problem.

One RFP-assigned mission was air drop. We ran flight tests with paratroopers but the tests were not very successful. There were some technical problems, not showstoppers in our book, but some Army leaders thought they ought to be. They were concerned about mission capability and the safety of their men. So were we, but we had to show them the fix, and bring them over to our side. It was time to bring in the consultants. We found some retired Army generals who had qualified as paratroopers. Once those guys put on the jump wings, they never took them off, no matter their rank or later career path. Instant rapport, instant credibility. Those officers, satisfied that we were on the right track, met with all of the senior area commanders, the folks who were most likely to be called in to testify before Congress about critical need and the solid C-17 solution. A great example if you will, of the "niche player" I mentioned back in the chapter on consultants.

A few months into 1993, the commander of Air Force Air Mobility suggested that the 747 and C-5 be revisited. I wonder, had those little devils at Boeing and Lockheed been re-testing the waters? In May, the deputy assistant secretary of defense for acquisitions and technology told John McDonnell, get on top of things, or else.

There were indeed many and continuing problems, focused not on safety of flight but on survival of the program. There had been long delays and unresolved cost issues. The "fixed-price" contract ran up against developmental realities; it turned out that while YC-15 successfully demonstrated new technology, scaling-up to the full-size C-17 was a whole different matter. The fixed-price didn't cover scaling-up, or change orders, or the rise in the cost of materials over the ever-growing length of the program. Whenever Congress withheld funds (they viewed that as an incentive!) or cut the number of airplanes, the cost of the program increased. There was plenty of blame to go around—for us, the Air Force program managers, and the DoD contracting officers—and issues that had to be resolved.

An inside source let me know that the deputy assistant secretary of defense for acquisition and technology was going to cancel our contract, give the program to Lockheed (wonder where he got that idea?), and bill us for something over 1 billion dollars worth of claims. My source also said the secretary saw John McDonnell as a drag on the program, therefore a part of the problem, and if he went away, we might be able to find some solutions. We would accept none of those options, although I don't think John knew about the hit on him personally. I sure didn't tell him.

After some ineffective efforts at long-distance negotiation, we arranged a meeting at the Pentagon with John McDonnell, the secretary, his general counsel, and me. The meeting got off to a great start: "I want you to understand just how serious this is," the secretary said. "You have so badly mismanaged the C-17 program that I'm seriously considering taking it away from McDonnell Douglas and giving it to Lockheed." We already knew that much; forewarned is forearmed.

I jumped in, put on my lawyer hat and said, "Mr. Secretary, just so you understand, I hope you will not bring that up again in any context. We lawfully won this competition against Lockheed. If you go through with such a grossly unfair and illegal action you will face the biggest lawsuit of the century." John gave me a strange look—I had not prepared him for this pre-emptive strike. "Now," I continued, "if we understand each other, let's see what we can work out."

Everyone agreed that the C-17 was going to be a great airplane. We admitted that some things had gotten out of control and needed to be remedied, and we all got down to business. There was a laundry list of items and in the next six months or so, over at least five more meetings with the secretary, we made some progress. However, toward the end of the year we were running out of steam. The secretary mentioned that he was replacing the officer responsible for the program on the Air Force side—call it, send a message that poor performance would not be tolerated. I suspected he was looking for a counter-offer. I already knew that he had a sacrificial John in mind, but once again, forewarned was forearmed. I had a different—better—idea and told him we were going to send a signal of our own and relieve the president of the company, the most responsible person on our side. Then both of us agreed to bring in new program managers on both sides; for our part, I nominated an individual well known to and admired by the secretary. The secretary said okay and that crisis, and the meeting, was over. There were more negotiations ahead, but so far so good.

Flying home on the company plane, I was expecting a few words of praise, like "That was pretty slick!" Instead, a coldly furious John McDonnell said, "We cannot work together." But he picked up the phone and called the company president, told him to hang around the office until the plane got back, they needed to talk. Then he told me, again, we can't work together, it was not good for the company and he passed the rest of the flight in angry silence.

Well, back at home, I had to tell Kate that I'd just been fired. She said, "Really fired?" and I said, well, 50-50, I don't know, but I'm going to lay low for a while and hope that John comes to his senses. Then, a few days later I found out that John didn't sack the president but asked him to announce his retirement. I called the secretary, "Does retirement send a signal?" He said "I'll have to get back to you on that." He called me back: "I don't give a damn how you characterize it, it's how I characterize it that counts and it counts as a change of management." John was home free, the program was saved, but there was still the matter of significant unresolved cost issues. More meetings were yet to come.

But not many, as it turned out. We soon made a proposal, which the secretary refused to accept. Almost at the end of the line, he gave us one more chance to work something out with the DoD contracts people, but he clearly was tired of the whole affair.

He gave us that grudging permission a couple of days before Christmas and, over the holiday period, with the secretary out of town, we had a long meeting with the Pentagon staffers who really made the system work—the contract coordinators and the financial and legal folks. We went over the items remaining on the list. It was amazing how quickly things moved along with people who knew what they were doing. The staffers agreed that some of the problems were caused by the government, which reduced the bill somewhat, and we agreed to settle the balance under a multi-year payment plan, where they would deduct something from each future progress payment, thus, no out of pocket. I'd say *that* was pretty slick. I kept my job, but John McDonnell didn't speak to me for almost a month.

As for the C-17: it did, indeed, prove to be an exceptional airplane. During flight-testing at Edwards Air Force Base, the C-17 set more world records than any other airlifter in history including payload to altitude, time-to-climb, and short-takeoff-and-landing marks; the C-17 took off in less than fourteen hundred feet, carried a payload of forty-four thousand pounds to altitude, and landed in less than fourteen hundred feet. The test program was honored by a visit from a female senator, known for her suspicion of things military and interest in feminist issues, and whose support we considered vital. It just so happened that the Air Force C-17 test pilot, who set many of those records, was Captain Pamela Melroy, who had recently returned from combat duty in the Middle East. There was no one better qualified to give the senator a tour of the airplane. Serendipity.

A footnote of sorts: Captain Melroy had earned her spurs flying KC-10s, including two hundred hours in combat and combat support missions in the first Gulf War, went to test pilot school in 1991, and was assigned as test pilot on the C-17 program. McDonnell Douglas all the way. She transferred to NASA in 1995, was shuttle pilot on missions in 2000 and 2002, and in 2007 now-Colonel Melroy was mission commander of STS-120 to the International Space Station and return.

The C-17 became operational in January 1995 and five months later was awarded the annual Collier Trophy "for the greatest achievement in aeronautics or astronautics in America." The C-17 also entered service with the Air Forces of the U.K., Canada, and Australia. The skeptics faded away when the planes began to demonstrate their true value. Perhaps the most telling example was during the 1999 Operation Allied Force in Yugoslavia, where fifty C-17s made up 7 percent of the airlift force and flew 61 percent of the sorties. There were 126 C-5s involved, managing 2.6 sorties per aircraft, to total 332. The C-17s averaged 24.6 sorties each, to a total of 1232.

Korea

The F-18 Korean adventure began in 1982 and, as noted before, was stumbling along until we sent in a manager who knew how to use the process—identify the decision maker, evaluators, and the influencers; discover the customer's Most Important Requirements (MIRS)—what does the customer *really* want?; hire the right agents; court the users—the Republic of Korea Air Force (RoKAF); assess the competition; anticipate their moves; and make an educated guess about their offer.

The competition was, basically, between our F-18 Hornet and the General Dynamics F-16 Fighting Falcon—then and forevermore the F-18's main adversary. In Korea, the F-16 had a significant head start, with the first of an earlier order of forty F-16C/Ds just coming into service. Korea was the first foreign operator of the F-16C/D. There was another contender, but it didn't count for much. Northrop, our partner on the F-18, was independently trying to market a sort-of new fighter, the F-20. This was an upgrade of a 1950s model, the F-5, and was in no way competitive. Northrop tried very hard to the extent that the CEO was publicly reprimanded for "his management style" and lost his job. On May 9, 1989, the *New York Times* reported rumors of some sort of attempted bribery.

The RoKAF was looking for 120 front-line fighters. What did they *really* want? They wanted to show their ugly neighbor to the north that they were armed and ready. They wanted to be seen as a world leader, be technologically advanced, and to come out from under the shadow of Japan. The government was looking for a co-production arrangement, bringing Korean industry in to the mix. No problem.

Of course, we had to get from here to there—from an understanding of the customer and the competition, to a winning offer, and go through a number of wickedly placed wickets. For example, General Dynamics had a well-organized information campaign, including frequent press releases and interviews. Many of the General Dynamics materials—a series of handouts called "Falcon Facts"—often contained errors. Well, a charge unanswered is a charge accepted; an error in the public file is a fact, forever. Our team in Seoul started a let's-set-the-record-straight newsletter, which was distributed to all the key media, government, and military officials, and published the day after an erroneous charge had been lodged. It was a neat trick, abetted by the fourteen-hour time difference between Seoul and St. Louis. They would send the suspect text to the home office in the Korean afternoon, which was breakfast time in St. Louis, giving our rapid-response team time to research and get the corrections back to Seoul the next morning to be quickly translated and passed around.

Our own information effort was more circumspect. Mr. Mac had a rule: never say anything bad about anyone's airplane. I don't know whether he wanted to protect the image of the industry or was just a proper gentleman. Well, Mr. Mac had passed on, his son John was now CEO, and I never heard John invoke the rule. I was happy to let my team say bad—but true—things about competitor's products. But not say them in public.

Our team kept a book full of reports, clippings, and observations about the F-16. They kept the book in St. Louis and from time to time would invite a Korean evaluator to fly over for a tour of the plant or some simulator time, whatever, and then take him off in a corner and show him the book. They would let him read it but not take it away. The take-away was another book, a running side-by-side comparison of specifications and features, F-18 versus F-16. On our side of the page, a list of all of the equipment and intangibles included in the asking price; on the other side, a list of all the things for which General Dynamics would charge extra.

The award was announced in December 1989. We won. General Dynamics had been so confident of a win that their chairman confronted our chairman with an angry accusation; we somehow had cheated because we brought in "agents" to work on the customer. The fact that General Dynamics also employed "agents" did not seem to matter.

Out team in Korea held a gala celebration in January. That was the high point of the campaign. We had been selected but were not yet under contract. In what was certainly one of the most boneheaded errors, ever, the F-18 program management decided that the job of marketing was finished, and it was time for the detail guys to take over, to negotiate the contract, and work the minutiae. Dealing with minutiae is important, but maintaining good relationships with the customer is even more so. Nonetheless, the sales team leader was transferred to head of sales for Europe.

New players drifted in from our side—"drifted" being the operative word. No senior McDonnell Douglas executive visited Korea for more than a year and the new program manager did not even visit the customer for six months. The center began to dissolve. There were changes among our top RoKAF supporters, including retirements and reassignments. There were rumors that the changes had been selectively managed to strip out our supporters. It wasn't long before no one knew anyone.

We heard reports that General Dynamics had been acting as if the competition was still underway, working on the National Security Adviser and the new RoKAF leaders. Our program management paid no mind. There were minutiae to be explored. We expected our offset partner, lined up for a significant portion of the work, to lend a hand. It turned out our partner didn't much care whether they were going to help produce F-18s or F-16s; they were in line to get about the same work share either way. Plus, they already knew something about the F-16, and the F-18 was *terra incognita*. When in doubt, stick with what you know.

I must note that the local U.S. Air Force command was not helpful to our cause; they were concerned, should war break out on the peninsula, about the problem of supporting an alien airplane—a Navy bird not in their inventory nor included in their supply and support chain. The fact that F-18s were well-supported, worldwide, did not mitigate their anxiety. This was not the first, nor the last, time that the "not in my backyard" mentality intruded on a competition.

After about a year of negotiations, dominated by unresolved minutiae, the Korean national security adviser announced a "re-evaluation." Three months later, he announced termination of the F-18 program (which had been seven years in the making) and announced a decision to buy another batch of F-16s.

Stories about huge "payments" swirled around the sudden re-evaluation, to the extent that the GAO launched an official investigation. Because the Korean government refused to cooperate, the results were inconclusive. Some months later, the national security adviser fled the country, the target of a corruption probe. In truth, a bunch of senior government officials, including the man who had been president from 1988–1993 and his predecessor, were later convicted of a variety of transgressions, including treason, mutiny, and corruption. One was sentenced to death, the other to more than twenty years in prison, but both were pardoned in 1998. I don't think any of the charges involved the switch from F-18 to F-16, but let the example stand as, well, symptomatic of the climate at the time.

Eventually, some good news came out of Korea: the 2002 buy of forty F-15Ks, the Korean version of the Strike Eagle. The sale was valued at 3.6 billion dollars. This time, there were no shenanigans and no competition. Thanks to heads-up ball by our team, the RFP actually specified that the offering was to be a "F-15-class aircraft." There were no F-15 class aircraft except the F-15. This time, the competition was so confident of another F-16 sale that they weren't paying any attention until it was too late. There were no other bidders.

Spain

It was 1982. We were trying to sell F-18s to Spain and not having a good time of it. Earlier, I noted some "offset" issues with this program. Those were important, but before we could even offer some offset we had to solve some real problems: the competition was coming on strong; there was a change in government and the new government seemed to be not very interested in the F-18. Our marketing team was fracturing with members fighting among themselves, and my "What's going on?" inquiries from St. Louis just brought a lot of whining. This was not a time to give up but rather a time to get creative.

I had recently attended a lecture by Edward De Bono and was impressed by his process for working through contentious issues by focusing on one aspect, and only that aspect, at a time. He called it his "Thinking Hats." The modus: everyone in the working session would pretend to don a hat of specific color, signifying the topic of the moment. A moderator would write down all comments on whiteboards spread around the room.

I decided to give it a try, although I changed his rules a bit; he had six "hats," I slimmed it down to three: black, white, and green. When you (metaphorically) put on the black hat, it was time to be critical, discuss difficulties and raise cautions about the issue at hand. No one could bring up any idea or mention any fact that did not fit under the black hat. This was not the time for arguing, just listing negatives. It was not the time for making suggestions or proposing solutions. Under the white hat, the group would come up with everything positive to say about the issue. With a green hat, everyone explored creative solutions. While this may seem a bit silly, it is a powerful tool for organizing thought and removing "ownership" from any aspect, good or bad. The negatives were no one's fault, they just were. The solutions belonged to the group.

I invited the team home from Spain for a two-day meeting. I explained the ground rules and started with the black hat: "Okay, we have two hours. I want everyone to come up with all of the negative things about the program and about the customer." This was an opportunity to vent. And vent they did.

The new government doesn't have any money. The new government is not interested in national defense. The political picture in Spain is totally fractured. Spain wants 110 percent of the program in offset. Military leaders won't come to our office in Madrid for a conference because it's too far from the air base. The F-16 is part of the NATO fighter program, why should Spain buy any other airplane? Why would Spain buy an American product when they are already a partner in the Eurofighter? We don't like the working hours—the whole town closes down from one to five, then works from five to eight and you can't get dinner until after nine. The Spanish have their own system for estimating what things should cost; we're never going to make the value argument to sell this airplane for the price we're asking. One of our agents is linked with gangsters.

After two hours of that sort of stuff, they were exhausted. It was time to go positive with a white hat. The new government may not be strong on defense but wants to be seen as independent, and will make an independent decision. The military may be naive, but there's a guy in the Ministry of Defense who really understands life-cycle costs. We were able to arrange a meeting with the new head of the Air Force. Our own senior executives love Madrid (they like the food, the ambiance) so we can get them to come help us with our customer. St. Louis University has a campus in Madrid that generates a lot of goodwill.

So after a while, the team begins to see, hey, it's not all negative. Now, what are we going to do about it? Under the green hat, the ideas began to flow: Invite the life-cycle-cost guy and the head of the Air Force to a briefing, which we arrange to hold at the air base. Get the CEO to fly in for a visit with the prime minister. Find out what influencers may have graduated from the University, get them involved. You get the idea. For my part, I gave the team the second day off. Re-grouped and re-energized, they went on to win. Spain bought 72 F-18s, joining Canada and Australia as foreign operators.

An aside comment: wouldn't it be wonderful if, say, political disputes could be handled this way? Fill a room with members of Congress and ask them to start by listing all of the things wrong with all of the proposals for, say, health care, without regard to political party of origin. Then list all of the good things about all of the proposals. Somewhere along the line, wouldn't you think that common sense might supplant partisanship and produce a bill with none of the bad and all of the good, more or less?

Switzerland

Switzerland wanted to buy a few dozen new fighters, something that could be parked in a man-made cave and fly out day or night, to leap above the mountains and do battle with an invading marauder. We knew that the F-18 was almost the ideal airplane for the mission.

Almost. Under a scenario posited by the Swiss, which included a timed segment from take-off to engagement, the F-18 didn't have enough power for the full-throttle leaping-over-mountains bit. That could be solved in one of two ways, by installing a "hot switch" that would provide a limited, short-term boost, or by developing and installing an Enhanced Performance Engine (EPE). In my book, the hot switch was a kludge, an instant loser (although it was favored by the head of the F-18 program). On the other hand, up-rating the engine for about 20 percent more thrust would take some money. The Navy wasn't interested in funding a new engine. They were happy with what they had.

It was time for a workaround. We whipped up a cost-sharing arrangement with the engine-maker. By our computations, the cost of development would be recouped with international sales of 150 EPEs, on which there would be a modest surcharge. The Navy would get a free ride, an EPE with no development cost. We took this plan to the Navy. The Navy was delighted. But the head of the F-18 program was furious. How dare I pull off a deal without his approval? I dare, because as the vice president for marketing, I didn't work for him. My job was to make sales, and this was the way to make this one.

The Swiss issued an RFP in 1985, and got seven takers: F-16, F-18, F-20, JAS 39 Gripen, Lavi, Mirage 2000, and Rafale. A couple of those—the Israeli Lavi and Northrop F-20—weren't real in-production airplanes, but what the heck, the Israelis and Northrop thought it was worth a try. There was a 1988 downselect to F-16 and F-18, followed by a four-week-long intensive fly-off. The F-18 won.

But wait a minute! The competition was re-opened two years later to accommodate the entry of a new player, the Mirage 2000-5. I guess the overly-cautious Swiss did not want to leave any stone unturned. The F-18 won.

Then local politics intruded. Switzerland is the world-model of a forever neutral nation, with no designs on its neighbors—ever; or at least since around the year 1674. In truth, we had to finesse semantics: the F-18 was widely known as a "Fighter/Attack" aircraft (F/A-18 in the U.S. Navy). The Swiss call their armed forces "Defense" forces; the possibility that they might ever attack a neighbor was unthinkable. They did not want to buy anything identified as an "attack" aircraft. It was easy enough to emphasize the "fighter" role, strip off some attack-specific equipment, and never mention the F/A designation. However, an anti-military faction tried to block the sale and the government put it up for public vote. That took another two years. We won. The public wanted to upgrade the Air Force.

Dealing with the Swiss was an education. Their acquisition process was detailed, discerning, and was conducted by a group of engineers. It was hard to sell them and it took longer than with most customers, but if the Swiss bought something every potential customer in the world knew that it must be pretty good. The contract was signed in 1993 with the first of thirty-four F-18s delivered in 1996.

CHAPTER 4
Finland

In 1989, as we were about finished (we thought!) with the F-18 sale to Switzerland, one of my guys wanted to make a pitch to Finland, which had expressed an interest in buying perhaps sixty new fighters. I thought he was crazy, Finland was too much under the thumb of the Soviet Union, and I told him not to waste time and money chasing a phantom. He went ahead anyway. That's what sometimes happens when you hire strong-willed, independent thinkers. He brought in a consultant, the former CEO of the Finnish national airline, Finnair, which operated an all-Douglas fleet. The man liked our products.

The Finns planned to split their buy, forty U.S. and twenty Russian airplanes. It turns out that they preferred American planes and were not enamored with the Soviet product, but wanted to keep on good terms with their restless neighbor. That latter concern began to dissipate when the Berlin Wall was torn down in November 1989, which perhaps opened the door to an all-U.S. buy.

Our small sales team struggled to get the Finnish Air Force to even take a look at a Hornet. There were several problems, although the first was easily solved: just like the Swiss, the Finns call their armed forces "Defense" forces. We were able to pitch the F-18 as a fighter, pure and simple, and offer the same somewhat stripped-down model that had been selected by Switzerland.

The second problem was more troublesome: the chief of staff of the Finnish Air Force was already committed to the General Dynamics F-16A, more or less the most basic model. The General Dynamics sales team and some F-16 champions in the Air Force hyped it as a special-model plane "flown by our National Guard for defense of the homeland" and therefore appropriately defensive. They also were feeding the Air Force, well, blatant misinformation about the F-18, such as the price being double the actual price and, oh, it wasn't much of a warrior, to boot.

In those days, before the advent of easy computer graphics and digital printers (and with a great deal of time and energy), the team prepared cut-and-paste exhibits from a range of briefing materials to be forwarded to the Finnish Ministry of Defense through the U.S. embassy. The team discovered, quite by accident, that our designated contact at the embassy—a colonel in the United States Air Force (USAF)—had been seen throwing our materials in the trash.

Were we supposed to put him on report through his chain of command? Well, no, because we knew that the corporate USAF didn't think much of this "Navy" airplane, the F-18. One of our guys, a retired Air Force colonel who wasn't on the F-18 program but was in-country to sell Apaches, took the embassy colonel out to lunch. It turns out he didn't have many friends. It's amazing what a simple gesture can sometimes achieve. Our next set of briefing materials made it through. It didn't seem to make much difference. Our ex-Finnair consultant tried to twist the right arms, but the Finns simply wouldn't include us in the competition.

I was busy enough with other programs and really didn't learn of the work-around until just at this point, after the team had spent several million dollars and we were being thrown out of contention. I was not happy—and that is a "mild understatement"—but, in one of those marvelous bits of accidental timing, I had a chance at that very moment to join one of the top Finnish decision-makers, an official in the Ministry of Defense, at a black-tie event in New York.

He was movie-star handsome and amiable; "Have you ever been to Finland?" he asked. I said no, and that the only thing I knew anything about was the English School. He said, "I went to the English School!"

The English School was started by my wife's aunt, Sister Claire Marie Weaver. In fact, she had been on a boat in transit when World War II broke out and was more or less trapped in Finland for the duration. When I told our guest, he became truly animated. "Sister Claire Marie is a national hero!" He opened up, told us about his childhood, about the school, about the impact Sister Claire Marie had on everyone. He also said, "Let me see what I can do about your airplane problem."

Sure enough, a few months later the Finnish Air Force invited us to bring in an F-18 briefing by the U.S. Navy's International Program Office. It did not go as well as we might have hoped, primarily because the Air Force chief of staff seemed not to have been listening. Clearly, here was a man with his mind made up. He wondered aloud, "Why did McDonnell Douglas abandon the single-engine heritage of the marvelous F-4 Phantom when it started building the Hornet?" Huh? The F-4 was a twin-engine aircraft. Conversely, he also warned that we would have trouble selling any twin-engine design to the Finns; a bit puzzling to us, because twin-engines in our airplane versus the F-16 single engine had been an advantage in every other head-to-head competition.

Not too long after the briefing, the chief of staff retired and was replaced by another officer who also had been at the briefing. He brought a different perspective: he was not a fan of the F-16 in general and the "air defense variant" in particular. He wanted a front-line fighter that was in the active inventory, not the militia. He also wanted to look at an all-American buy, which would simplify procurement and operations, and bring overall cost down (something that could work in our favor; there was no question, the F-18 would be more expensive than the F-16A under any scenario).

One of the team, a former naval aviator of Finnish descent and our on-scene coordinator, tried to get the new chief of staff to take a ride in a Hornet, but the general did not want to show favoritism. Our man then learned that the previous chief of staff—Mr. F-16A—would be at an upcoming air show and went there to offer him a demonstration in a two-seat version. Well, okay, the retired general said, why not? A free ride in any top-line military aircraft is always interesting.

Later, when the new chief of staff heard about this, he said, "Okay, I'll take a ride also." But we gave him more than just a "ride." We flew him to a USN aircraft carrier out in the Atlantic and treated him to a weapons demonstration and eight launch-and-recovery cycles. When he climbed down from the cockpit he was as excited as a child, pounding his fists, marveling at the power and might he had just experienced. I think any lingering interest in the F-16A had just evaporated.

Some months later, at another big event—the 1991 Paris Air Show—Finnish Minister of Defense Elizabeth Rehn requested a briefing on the F-18. We had a list of the questions she wanted answered: How long would the planes last? Why use a Navy airplane for land-based defense? We had the questions because the Finnish Air Force had passed them to us under the table, so to speak, and told us how Minister Rehn liked her briefings. We suspected, but couldn't know for certain, that the Air Force was now on our side.

We had charts set up with all of the answers. Minister Rehn seemed properly interested, nodding her head at the right points, but did seem a bit preoccupied. With the briefing over, she turned to her nominal host and said, "We have one more item to discuss." The Finns were building some training aircraft, the L-90 Redigo, and to this point had only sold ten of them, all to the Finnish Air Force. "We'd really like your help with this program," she said. Well, we had a standard follow-up question after any program loss ("Did you do everything you could have or should have done to win?") and the host did not hesitate a moment in agreeing to help. In fact, we bought ten L-90s, thus doubling the production run. I don't remember what we did with them, maybe sold them to Eritrea and Mexico (which together seem to have acquired ten L-90s).

At the same air show, I believe, British Aerospace was trying to sell Minister Rehn some armed trainers, Hawks, and offered her a demonstration flight. She declined. Apparently British Aerospace was not aware that she already had taken a ride in the backseat of a Finnish Air Force Hawk, which crashed, and she was almost killed. There is a lesson here.

In December 1991, the Soviet Union imploded and with it went the last vestige of politically-driven interest in buying Russian airplanes. The Finns would buy all the fighters from the Americans, and they were not interested in splitting the buy. The issue was decided by a small group that included the "English School" graduate I had dinner with in New York and the group was aided in their deliberations by our consultant. The Ministry of Defense invited the head of our European sales office to come to Helsinki and pick up the RFP.

On a very cold winter day, our man and an associate arrived at the MoD expecting a quick office call, but found themselves in a meeting at the MoD guest house with the two top procurement officials. The Finns made a sharp point: including McDonnell Douglas in the competition had been the subject of much discussion. It had been a difficult decision, and they wanted assurances that McDonnell Douglas would take the RFP very seriously. That said, our team was invited to join them in the sauna (about as common in Finland as a kitchen) for an evening of heavy drinking, joke telling, and sweating, where they had the distinct impression that the Finns had already decided on the F-18, but couldn't say so straight out.

In May 1992, the Ministry of Defense was pleased to accept our offer: 64 F-18s, price 3 billion dollars. At the contract signing, the Finnish Air Force chief of staff said that we had won for a very basic but hardly simple reason: "We are going to have these airplanes for more than thirty years. We absolutely trust your team and company to partner with us, to ensure that all will go well." That, dear reader, is what it is all about.

Our on-scene coordinator soon got to move from Helsinki to St. Louis as Finland F-18 program manager. The first aircraft was delivered in 1995. As of this writing, the Finnish Air Force operates 63 F-18s, having lost one in an accident. I've heard that they have re-installed the "attack-related" equipment that had been stripped out in the "conversion" from F/A to F-18.

If You Can't Lick 'Em, Join 'Em

A big opportunity popped up shortly after I took over the Helicopter Company: the U.K. was in the market for a top-of-the-line attack helicopter, and the competitors were lining up. According to my best sources, they included three British companies (helicopter builder Westland, BAE teamed with the German/French Eurocopter, and GEC-Marconi paired with the American Bell Helicopter Textron), one all-American team (Boeing and Sikorsky), and the Italian Augusta. My sense was that the British would favor a home-grown product, but the only purely British play on that list was the financially-shaky and product-poor Westland, which did not stand much of a chance. Ah, but what if they were assisted by the company that built the attack helicopter poster boy, the Apache?

I suggested a deal: Westland would be the prime contractor; we would license the Apache to them for sale in the U.K., and, as a sub-contractor, supply the parts from which they would assemble finished helicopters. I was feeling pretty proud of myself when I sketched out the details for the McDonnell Douglas executive committee, but they were aghast. Someone said, "We send you out to be president of the Helicopter Company, to keep it from falling off a cliff, and the first thing you do is give away the store?" I said, "hey guys, whatever title you pay me for, I'm a salesman first, and this is a sale that will bring in money." I reminded them that Northrop was making a bundle of money as subcontractor on the F-18 without the burden of risk that we had to carry.

Someone said, "You need a bigger partner, they can't possibly handle such a large program." I said, in my judgment, Westland will be the political favorite and the key to getting the contract. We can make the deal work. I was right. Our team got the contract, 67 Apaches, priced out at about 5 billion dollars, of which we got the larger portion. The concern over Westland's financial situation was soon erased when the much-larger company GKN, which had a small stake in Westland, heard about our deal, increased their holding, and just before the contract was issued turned Westland into a wholly-owned subsidiary.

If I ever needed validation of my strategy, I got it at the press conference where the minister of defense stood up and said, "I've come to announce the results of the U.K. attack helicopter competition. We are awarding the contract to the Westland Apache, powered by Rolls Royce engines and armed with Shorts Starstreak missiles." U.K. all the way. Not a whisper about McDonnell Douglas. I didn't mind.

On another, later, U.K. contest, someone said that we needed a British partner and suggested a plausible company. His boss said, let me check it out and called an English friend, a commodore, who was involved in acquisition and who said it wasn't a good idea because that company is high cost and they have a lot of risk. So we didn't do it. It turns out, that was a mistake and the biggest mistake was in asking the wrong person the wrong question. We should have asked the prime minister—the man who made the decision—a more open-ended question, "Who should we team with to enhance the opportunity for both companies?" Chalk this one up to an engineer, not a marketer, working the problem. Engineers are more comfortable dealing with folks in the acquisition community, who are largely also engineers. I'd like to say that marketers are, well, fearless, and maybe they are, but in this case, they certainly know how to ask the right questions of themselves (starting with the obvious, "Who is the ultimate decision maker?") before they go poking around.

Let me finish by adding one small "partnership" story from France. I had at least four meetings with Serge Dassault, head of Dassault Aviation, to see if we could create a partnership to build a new fighter. McDonnell

Douglas would provide some technology, he had extensive production experience, facilities, and a worldwide sales team. Together, I thought, we could capture a large part of the world market that did not look favorably on dealing with the Americans. Sort of like the Apache deal with Westland. Serge was quite interested, the advantages were obvious, but he had some concerns—mostly, money and time. It would cost a bundle and take forever and get in the way of improving his current product line. I offered the F-18 planform to jump-start development to get a new and better fighter to market quickly and don't worry about upgrades of the old Dassault models. Serge liked the idea, a lot. Apparently the French government, which owned almost half of the company, did not. Co-develop with the Americans? Never!

Jobs Now

In July 1992, Representative Howard Berman (D-CA) called a news conference. "I think I am watching," he said, "the most sophisticated and far-reaching campaign to promote a sophisticated arms transfer that I've ever seen since I've come to Congress." He was talking about the McDonnell Douglas campaign to promote the sale of 72 F-15s to Saudi Arabia. He did not mean this as a compliment, but we certainly took it as high praise, indeed, for what was the first major campaign launched under the new Acquire Business Process. However, this was a campaign with a twist: we didn't have to sell the customer, he wanted the buy. We had to sell the idea to anyone in the United States who had the power to approve or block the deal, and, in so doing, I believe we set the gold standard for grass-roots marketing.

This is a tale of success, frustration, and geopolitics that began with the 1978 sale of F-15s to Saudi Arabia (in which I acted as strategic coordinator). At the time, because of Congressional concern for Israeli sensitivities—the Saudis and Israelis were both allies of the United States but not of each other—the aircraft were somewhat de-tuned, range was reduced, and the Saudis were limited to 60 aircraft in-country at any given moment.

In 1985, the Saudis wanted to buy some more but the sale was blocked by objections from the American Israel Public Affairs Committee (AIPAC), a heavy donor to political campaigns. Here's a little story: AIPAC was assembling signatures of senators opposed to the sale. Get enough names and there would clearly be no point in trying to move legislation. I had a strong promise from one key member, "You can count on my support." Well, his name showed up on the list and in a rage I flew to D.C. and confronted him in his office: "You son-of-a-bitch, they offer you something like a million dollars for your next campaign and you cave." "Oh, Tommy, Tommy, Tommy. They are much more sophisticated than that. They said they would give a million dollars to my opponent if I didn't sign. So they got what they wanted and it didn't cost them a cent!"

The Saudis would have preferred F-15s but didn't seem to care all that much. They had a back-up plan, to buy some fighters from the French or the British. There had been some preliminary talks and the French seem to have been favored, at that point, should the Americans back away. When the White House told the Saudis that the president could not support the F-15 deal, a Saudi official told the French and British aerospace marketing guys. The British marketing guy (who probably knew that his team was behind but knew how the game should be played) enlisted the aid of Prime Minister Margaret Thatcher, who engaged in one-on-one negotiations with the Crown Prince and soon enough was presenting herself, properly decked out in hat and gloves, to the Saudi king.

Under the largest export program in British history, the Saudis bought 72 Tornado fighters, 90 Hawk advanced trainers, 30 Pilatus basic trainers, helicopters, some minesweepers, communications systems, and assistance with construction of air and naval bases. Dubbed "Al Yamamah" (The Dove), the project eventually pumped 28 billion dollars (some accounts say 80 billion dollars; I won't take sides) into the European economy. Israeli security was no different than it would have been had the Saudis bought F-15s. AIPAC gained nothing and American workers lost. This was a lesson for the future.

That future arrived in 1991. F-15 production was winding down, cleaning up the last U.S. Air Force F-15s under contract for delivery by 1994. There was nothing in the order book. In the euphoria over the success of the 1990 Gulf War, as after every war, national priorities shifted from defense to domestic spending and there would be no additional F-15 orders from the U.S. Air Force. Layoffs would begin in summer 1992, with skilled workers drifting off to other employment. Once that process begins, it's like a dying man whose organs begin to shut down one at a time. He is still alive, but with little to offer and not for long.

The only chance to keep the line open was a big international sale. We knew that the Saudi long-range plan had always been to add more F-15s and in October 1991 they told us they were ready. The political climate had warmed a bit, thanks to Saudi participation in the Gulf War, during which—without controversy—the U.S. Air Force transferred 24 F-15s to the Saudis and in which two Saudi F-15s shot down two Iraqi MiGs. In fact, at the same time, while I was working at the Helicopter Company, I sold them 12 AH-64 Apaches without difficulty. Perhaps the Saudis were no longer radioactive.

The first public notice of interest in a new F-15 buy came out of the Dubai Airshow, November 1991, and yes, opposition quickly arose. However, the rationale was not directly focused on Israeli security. Two senators prepared a letter, addressed to President George H. W. Bush and endorsed by 67 of their colleagues, expressing concern over fresh arms sales in the Middle East. Congressman Howard Berman prepared a similar letter, signed by 238 members. The Saudi Ambassador to the United States, Prince Bandar bin Sultan, made a quick phone call to one of the senators, suggesting that his country was not actively pursuing a sale "at this time." The legislators pulled back and we had time to get our ducks in a row.

But we were now dealing with an additional and broader issue: in summer 1991, the Bush administration, supported by many members of Congress, had called on the international community to restrain arms sales to the Middle East. The United States, Great Britain, Peoples Republic of China, France, and the Commonwealth of Independent States (read: Russia) all agreed to think about the problem. That was about as far as it got with most, thinking. There was too much money at stake.

At McDonnell Douglas, we firmly believed that the F-15 sale was in the national security interest of the United States and would not create instability in the Middle East nor disturb the ongoing Middle East peace process. Our challenge was to convince a majority in the Congress and the

Bush administration of that truth. Point man? Jim Caldwell, a retired Marine, which was (and still is) the most energized of the armed forces, and Jim had (and still has) an uncommon grasp of domestic and international politics and policies.

Our strategy? To create an overwhelming groundswell of public support and to defuse any and all efforts to deny the sale. Our tactics? We brought in consultants, experts in grass roots public relations or people who were well-connected with the communities we needed to reach. In December, we created a coalition of F-15 contractors: McDonnell Douglas, Hughes Aircraft, and United Technologies (parent of engine maker Pratt & Whitney), which were soon joined by other suppliers, unions, and business groups. Three other major aerospace companies, Martin Marietta, Northrop, and General Electric, came aboard to join in a letter to President Bush. The March 30, 1992, *St. Louis Post-Dispatch* called it "Group Therapy."

We agreed on three basic messages for the campaign: First, forty thousand jobs were at stake and we should keep those jobs in the United States. Those jobs were spread over more than 3,500 suppliers in 47 states and 346 congressional districts. (Opponents challenged our numbers but we were confident in the methodology and no one was able to prove us wrong.)

Second, the sale would add perhaps 13 billion dollars to the U.S. economy, and not just from the airplanes. The package included twenty-four spare engines; navigation and targeting pods; nine hundred maverick missiles; and a couple of thousand other missiles and bombs, support, spares, training. Our share was just under 5 billion dollars and worth fighting for.

Third, we believed that the security of Israel would not be compromised and stability in the Middle East would be increased. The United States could impose restrictions of the use of the aircraft (enforced by controlling support and spare parts), while European companies would not likely do so and the Russians, never.

We tested the messages before focus groups and by surveys. Emphasis on "jobs" tested the highest, especially when paired with the fact that if we couldn't deliver, the Saudis would once again buy from European competitors.

In that event, who might benefit? Everyone was waiting in line: the French had lost out to the British in the 1986 Al Yamamah competition.

They were displaying remarkable patience, ready to grab the next opportunity. Serge Dassault told the June 17, 1991, edition of *Defense News* that, "Time is working for us. In three to four years, there will not be any U.S. fighters cleared for export in production."

The European Fighter Aircraft was on the drawing boards, but the governing coalition was teetering. There were reports that Germany was about to pull out and the February 10, 1992, *Defense News* reported that the Saudis would come in as a partner should the F-15 deal fail.

The British were winding down their Tornado production and a new sale would be most welcome. In April 1992, British Prime Minister John Major, up for re-election, was happy to announce that the Saudis had just sent 2.6 billion dollars to Great Britain, a progress payment on Al Yamamah. A British Aerospace official said he expected a new order for Tornados "within weeks." (As an election ploy, it didn't work; as noted on page 100, John Major's Conservative Party was defeated by the Labor Party of Tony Blair.)

Circling around just outside the center were Sweden, China (building copies of Cold War-era Soviet aircraft and making efforts to produce a home-grown fighter), and always, Russia. The November 8–9, 1991, *Wall Street Journal Europe*, in covering the Dubai Airshow, noted that the Russians announced a deal to sell one hundred MiG 29s to Iran. "We're ready to sell them everything," a Russian official said. "Everything. Look, let's face it: We are broke."

If the Europeans gain, we lose. The fact that some of those Europeans, most in fact, had higher unemployment rates than the United States and needed the jobs was unfortunate but, in this context, not compelling. We had our own parking lots to fill. We created the slogan, "U.S. Jobs Now."

There was public uncertainty about the long-term loyalty and stability of the regimes in the Middle East. Was it wise to sell advanced weapons systems to nations that might change interests? Memories of the Iranian revolution were fresh enough. We held focus group sessions of potential supporters (veteran's groups, conservative republicans, affected employees) and probable opponents (politically active liberals). We found that most viewed Saudi Arabia as an "an ally of convenience" rather than a "trusted" ally, but few regarded the sale as a direct threat to Israel.

In a poll of coalition partners and organization leaders, seventy-four percent favored the title "U.S. Jobs Now." Those who objected said it sounded like a "job handout program"; was "corny"; and smelled of "popularism, liberalism, and unionism." However, most respondents would be willing to take some positive action—write letters, attend meetings, make phone calls to Congress, send a letter to an editor.

We had participation or support from six labor unions with 5 million union members and more than sixty civic, business, and political organizations. Together, they sent 20,000 letters and made 10,000 phone calls to congressional and administration offices. We held more than five hundred briefings for members of Congress and their staffs. We held rallies in twenty-one of the larger affected cities. We distributed 150,000 bumper stickers, 150,000 brochures, 25,000 T-shirts, 10,000 branded pens, 50,000 memo and note pads, and we passed out 2,700 videotapes.

The April 18, 1992 edition of *National Journal* called the video "slickly produced" with a "stirring sound track." Well, why not? We wanted to emphasize that about 12 percent of U.S. aerospace workers lost their jobs in the last year, buried in a blizzard of pink slips following the end of the Cold War. Some people respond to cool intellectual messages in brochures and point papers; some people like a bit of drama. The video provided drama. "Our military aerospace industry has made history. Will it be here to shape the future?"

Most of our advertising was concentrated in the National Capital Area: *Congressional Quarterly, National Review, Roll Call, Washington Post,* and *Washington Times.* We followed the classic advertising formula: find the unique selling proposition and repeat, repeat, repeat. The most effective print ad was a full page filled with the names of all the cities involved, with the headline: "Who needs 40,000 highly skilled jobs? We do!"

We were up and down every month, winning or losing. There was a big chart on the wall showing all of the forces that were with us or against us. The "against" column was daunting. The assistant secretary of state for politico-military affairs said, "It is not the United States Government's policy to sell arms in order to maintain our own defense industrial base." Other senior officials at the State Department and the Department of Defense went on record in the February 26, 1992 edition of the *New York Times*

(Eric Schmitt, "White House, Eager and Anxious, Want to Sell 72 F-15s to Saudis") as opposed to the sale because it would imperil the Middle East peace talks. On March 8, a *New York Times* editorial asked, "Why should the U.S. stimulate competition to sell advanced fighter planes to the Middle East when it could instead promote international cooperation to shut down arms sales?" In April, the *New York Times* took the *Los Angeles Times* to task for reporting—inaccurately—that the Saudis had been trafficking in illegal transfers of U.S. arms to Iraq and Syria. The Los Angeles paper got the information from an anonymous source who wanted to poison the Saudi well (Leslie H. Gelb, "Foreign Affairs; Trashing the Saudis," *New York Times*, 24 April 1992). In the same edition of the paper, the *New York Times* reported that President Bush had already "decided against asking Congress to approve the F-15 sale this year." That did not mean that the program might not be approved at some time in the future, but this was of scant comfort. Next year? There would be no viable F-15 production line.

We did have some powerful friends on our side. In March, Secretary of Defense Dick Cheney and Joint Chiefs Chairman General Colin Powell testified in support. House Majority Leader Dick Gephardt (whose district was, well, St. Louis) suggested that the sale would be approved if the administration announced support. There was no word from the White House. Israeli elections were coming up in June, and the administration was avoiding any actions that might seem to be an effort to influence the outcome. That election was held June 24. No word from the White House. In July, CEO John McDonnell sent an urgent letter to President Bush stating, "Layoffs have begun." No response. In the middle of August, newly elected Israeli Prime Minister Yitzhak Rabin came to the United States and told AIPAC that he would not work to block the sale. Democratic presidential nominee Bill Clinton announced his support. Rabin also visited President Bush at his Kennebunkport summer home. One of Bush's neighbors (a guest at the reception) heard Rabin tell the president that he did not oppose the sale. The neighbor called Jim Caldwell and told him the gist of the conversation. Nothing heard from the White House.

Some folks in McDonnell Douglas senior management had given up and wanted to pull the Saudi F-15 sales team, find something useful for them to do. I wasn't ready to give up, much to the irritation of some in senior management. As noted earlier in this narrative, relationships are everything. I pulled in some markers and some well-connected friends delivered a pointed message to the White House: with the election so much in doubt, does the president really want to write off Missouri along with a good chunk of 346 congressional districts? The Israelis are not worried. What else does he need to know?

No word from the White House.

And then on September 10, Jim Baker, who had just left the job of secretary of state to handle the president's re-election campaign, tracked John McDonnell to a meeting in California and told him to get back to St. Louis, pronto. The president was going to drop by the plant at 3:30 the next day. The visit wasn't on the president's published schedule. We didn't care. When the Secret Service moved the visit up one hour (as the Secret Service often is wont to do with only a couple of hours notice), we didn't care. We managed to have 35,000 people assembled for the occasion including, of course, John McDonnell and Missouri Governor John Ashcroft. A casually-dressed Senator Kit Bond (short sleeves and a baseball cap) introduced Bush.

We didn't know what the president was going to say until he said it: "I have decided to notify Congress to sell up to seventy-two of your F-15s to the country of Saudi Arabia." The president acknowledged recent hard times and that he knew the F-15 line could be winding down. He wanted to keep Americans at work, of course, but was concerned about stability in the Middle East and wanted both to ensure that Israel would maintain a qualitative edge while meeting the legitimate defense needs of another ally, Saudi Arabia. "I have worked on this issue personally," he said, "touching every base, and I am now satisfied . . . that we can and, indeed, must, for our own interests, go forward with this sale." He vowed to fight any congressional effort to cancel the deal.

The president also said, "The military technology that you produce is the finest in the world." I should note that he said the same thing when he announced the sale of 150 Fort Worth-based General Dynamic's F-16s to Taiwan the week before. This was another politically-controversial sale, one that had been on hold for more than a decade. It was no surprise that a politically-challenged president from Texas would want to preserve jobs in

Texas. I might also note that, as reported in the September 12, 1992, issue of the *New York Times* (Andrew Rosenthal, "Bush Plans Sale of 72 F-15 Planes to Saudi Arabia," p. 1), the White House insisted (in connection with both announcements) that the forthcoming election was not a consideration. Oh, perhaps. We didn't care. We just wondered why it took so long.

The administration made formal notification to the Congress three days later. Congress had 30 days in which to block the sale. Some members made half-hearted moves in that direction. Senator James M. Jeffords, R-VT was quoted in the same September 12 *New York Times* article as saying, "If we're serious about the proliferation of weapons in the Middle East, we ought not be out front selling to Saudi Arabia. We ought to be leaning extremely hard on the British and anyone else not to sell to them." Forty members of the House supported a "don't sell" resolution offered by Congressman Berman, a significant reduction from the 238 who had signed on to his earlier letter of concern. Another senator tried to impose a six-month delay. Congress adjourned without taking any action, which was, in effect, action enough.

The F-15 line stayed open and in the years since, McDonnell Douglas built 163 F-15s, enjoying 11 billion dollars of revenue. As of 2007, the program still has eighteen hundred people working on it.

Some of those 163 were purchased by Israel, which had been planning an F-16 buy. Some folks at the company wanted to push the F-18. I didn't think that made any sense. Nothing had changed in the Israeli acquisition strategy since my 1978 meeting with the Air Force chief of staff, when he explained that fiscal realities mandated a high-low force structure. Yes, the F-18 had more capability than the F-16, but not enough to fit the definition of "high" and the cost difference was simply too much. To go head-to-head would be a waste of time and money. However, I didn't want to leave the field without doing battle. How might we convince them it was time to take another look at the F-15?

I brought in some consultants, boned up on regional issues, did a threat analysis, made a trip to Tel Aviv, and posed some mission scenarios that could not be accomplished by even the latest, improved version of the F-16. The Israelis agreed. It was time for a purchase at the high end. They bought twenty-five of the latest model of the F-15, the Strike Eagle. It was modified to meet Israeli requirements and dubbed the F-15I.

The Saudis and Israelis had a gentleman's agreement: neither openly acknowledged that the other was buying more-or-less the same airplane (there were differences in installed equipment). However, at one point the Saudis were running out of money and had to defer delivery, so some Saudi airplanes were re-branded and sent to Israel. Later, some Israeli-destined airplanes were modified and sent to Saudi Arabia. Once, a Saudi official toured the assembly line, with the Israeli F-15I on one side and the Saudi F-15S across the aisle on the other side. The official pretended not to notice. Sometimes, you can't tell the players without a program.

Saudia Airlines

Early in 1993, as some F-15 contract issues were being resolved, the Saudis launched a contest to supply new commercial airliners to Saudia, the national airline, and to Royal Flight, the king's private air fleet. Saudia was ready to replace Boeing 747-200s, Lockheed L1011s, Airbus A300-600s, and Boeing 737-200s, all of which dated from the 1970s and early 1980s. Saudia also wanted to add four freighters, and Royal Flight wanted to replace two VIP-configured L1011s. We snagged the Royal Flight deal right away, largely because the company had two unsold MD-11s sitting around and made a good deal.

Aerospace industry smart money was betting that the 747-200s would be replaced by 747-400s, the Airbus A300s and the 737-200s by the new A320, and Boeing and McDonnell Douglas would wrestle over the L1011s, pitting the brand new Boeing 777 against the MD-11. We were happy, of course, to compete for the L1011 replacements, but with the Douglas commercial order book getting thin and with a competitive product right in the neighborhood of the A320 (the new MD-90, just then in development) the company wasn't going let the A300/B737 replacement go to someone else by default. We decided it was time to call in some big guns.

Thus, with a nudge from Douglas and energized by the prospects of layoffs (the "Jobs Now" concept in automatic), President Bill Clinton made a phone call to the Saudi King Fahd bin Abdul Aziz, extolling the virtues of American products, especially airplanes. In the meantime, ah, the

French! The French and Saudi governments got into a dispute over an intelligence matter, or it may have been over increasing French contacts with Saudi mortal enemy Iraq, or both, but the French would not back down, and the Saudis effectively put the A320 on permanent hold. (Or at least, semi-permanent. In November 2007, they finally signed an order for twenty-two A320s.)

I have to say, playing the political card may have been a tactical error. When the palace let it be known that Saudia would enter into discussions with Boeing and Douglas, the director general of the airline made it clear that he was more comfortable with Boeing (which had played hands-off in the politicking) and that he resented the intrusion of politics into a private transaction. Sort of private, that is. We all knew the airline did not have a free hand in the selection, and that they would do whatever the government told them to do, but they wanted, at least, to appear independent. A neat trick, since the chairman of the airline was also the Saudi minister of defense and aviation, who had to approve any deal.

So, it was to be open competition with Boeing to replace both the mid-size L1011 (Boeing 777 against our MD-11) and the smaller A300/B737 aircraft for which Boeing offered a newer model 737 ("Easy transition for the pilots") against our MD-90 (which we pitched as "Larger, more modern, with a state-of-the-art glass cockpit"). Both companies had teams living in Riyadh, meeting with the same officials, but information was held remarkably close. Over one period, our team filed almost daily reports with me and the home office (with Saudi players identified by codenames: Torpedo, Gatekeeper, Big Boss, Spaceman, Counselor, and Youngster. Not very sophisticated, I know. I suspect that any rational spy could guess, as least, the identity of "Big Boss" who lived in the "Big House.") There were plenty of rumors; we heard that one buy would be a mix of fifteen MD-11 and twelve 777. A senior adviser to Saudia warned, "Why would you buy two types of airplane for the same job?" A general in the Royal Saudi Air Force put in a word for the MD-11. It would make a good tanker, he said, and fleet commonality would be a good thing. Someone planted an article in a local newspaper noting the worldwide acceptance of the 737 and that the MD-90 "is not well-accepted in the region." Of course, the MD-90 was not yet operational, but why left facts get in the way of a good argument?

While we had a deal with Royal Flight, there were some bumps. We got word that one official objected to paying us a "management fee" to handle the MD-11 VIP conversion (which would be accomplished by a European company) so we changed the offer to read "engineering, coordination, and integration fee." It really should have said that in the first place.

The Saudia negotiations turned serious in March 1994. There were issues of financing, loan guarantees, and price. There were also payment issues. I heard of one scheme in which Saudi oil production would be increased from 8 million barrels per day (mbpd) to 9.5 mbpd, and the difference would be transferred to whoever as a credit at the then-current rate of $11.99 per barrel.

The Saudis had an interesting negotiating tactic, something along the lines of, "You are a long way from Boeing" and not much else. At one point, discussions moved to our Washington office. You are no doubt familiar with the ancient adage, "When in Rome, do as the Romans do." I offer a modern twist: when you invite the Saudis to your country, be prepared to treat them as if they were in their own country. To promote a sale, do whatever is appropriate to accommodate the customer. In this case, that meant renting prayer rugs. (Yes, there was a rental service.)

Well, the discussions broke down and the Saudis said they were leaving, so one of our staff exercised initiative and took the prayer rugs back. Perhaps he wanted to avoid rolling into the next rental period. However, while he was gone, and after some mild arm-twisting, the Saudis relented and the negotiations resumed. Our rug man returned to the office, saw what was happening, turned around and rushed back to the rental service, returning with the rugs just in time for the next session of prayer.

Finally, more than a year after the competition began, the deal was done but not to be revealed until a grand event in Riyadh, a gala celebrating Saudia's 50th anniversary. This was important enough for John McDonnell to make the trip to get the news. The news was mixed. Saudia bought twenty-three 777s and five 747-400s from Boeing; and twenty-nine MD-90, four MD-11 freighters, and the two MD-11 VIP-configured for Royal Flight. While the dollar value of the Boeing deal was higher, I would call it a fairly split decision. At around 2 billion dollars, it was one of the largest firm dollar orders in the history of Douglas Aircraft.

Joint Defense Attack Munitions

In 1993, five companies worked up proposals for a new strap-on guidance kit for standard bombs. The goal of the Joint Direct Attack Munitions (JDAM) program was simple: to greatly improve the accuracy and effectiveness of gravity bombs by adding a GPS-guided steering system. The "Joint" meant JDAM should work with a wide variety of Air Force and Navy aircraft, and on bombs ranging in size from 250 to 2,000 pounds (with the guts of the system mounted in different size adapters). The idea grew out of the 1991 Gulf War, when weather and battlefield obscurants seriously impeded the ability of pilots to get bombs on designated targets. Suppose, someone mused, we had a bomb that could be programmed to steer to a target—with settings put in before takeoff or even while in flight? Suppose, indeed. When a bomb is released, it is moving at the speed of the airplane, and if released at the optimum attitude and altitude, a steered bomb could be effective when launched as far as fifteen miles out from the target.

In April 1994, the original bidders were down-selected to two: Lockheed Martin and McDonnell Douglas, each given an eighteen-month engineering and manufacturing development (EMD) contract to see what they might come up with. To this point, JDAM was being handled just like any other big-dollar acquisition. The RFP for this rather basic program, not technologically challenging, in the traditional sense, ran to one thousand pages. Bidder's proposals had averaged around five thousand pages and ours was about six thousand.

JDAM—basically a low-cost, high-volume product—was quite different from the sort of high-tech, high-cost, slow-rate programs on which the company usually bid. The baseline product of the missile division, competing for JDAM, was Harpoon, then pushing 1 million dollars each, selling at the most two hundred units a year. The government planned to buy 40,000 JDAM kits, didn't want to pay more than 40,000 dollars each, and take delivery at a rate of 5,000 units a year.

The folks working the program had been among the first to be trained in the New Business Activity. There was some agonizing—bid or no bid. Someone in DoD had estimated that the units would probably cost 68,000 dollars, so shooting for 40,000 dollars would certainly be a challenge. However, there were some things that favored a try, especially the successful track record with Harpoon. Perhaps some components of the guidance system could be adapted to JDAM. A contract for 40,000 units at, say, 40,000 dollars came out to 1.6 billion dollars. It was worth a shot.

And then came a strange and totally unexpected shift in focus. Just after the award of the EMD contracts, DoD decided to drop the "conventional" development program in favor of a radical approach; a trial-run at streamlining the acquisition of military equipment. The winners were told more about "what" to do—design and build a kit to meet the mission goals—than about "how" to do it. There were six immutable requirements: the cost must be below 40,000 dollars; the accuracy must be within thirteen meters of the target in the most adverse weather; there must be compatibility with designated aircraft (such as B-2, F/A-18C/D) and with specified warheads; JDAM must be suitable for use on aircraft carriers; and it must be re-targetable while in-flight. Other than that, contractors could start with a clean slate and a clear mind.

This approach had been in gestation for some time, suggested in several pieces of legislation and a 1994 DoD memo, "Specifications and Standards—A New Way of Doing Business." It was, in truth, something that had been pushed for many years by those of us in the industry who knew there had to be a better way, to get away from the government's running suspicion, fear of being cheated, and certain knowledge that only they knew what was best. Streamline, simplify, cut out the stifling oversight and the regulations that added time and cost to any procurement without adding value or ensuring success, in favor of a more commercial approach. This was an idea whose time had come—sort of. There were people in DoD who simply didn't believe it could work but were willing to try.

Let's compare typical DoD practices with the commercial approach:

- DoD would say, "give us the best possible product." Commercial: give us an affordable product that will do the job.
- The typical DoD contract offered agonizing detail on "how-to-do it," the commercial approach, "here is what we want it to do, find a way."

- On pricing, DoD favored "cost plus" contracts: "tell us what it costs and add x percent for your profit." A bookkeeping headache, with no built-in incentive for the contractor to keep costs low. For a simple example, if the cost was 100,000 dollars and the contractor got, say, 10 percent, the final cost to DoD was 110,000 dollars. If the contractor might find a way to trim the cost, his profit went down. The "plus" on a slimmed-down 80,000 dollar buy was only 8,000 dollars. The contractor lost 2,000 dollars trying to save the government some money. With commercial pricing, there was a fixed unit price from first item to last, which included a modest profit.
- The DoD approach was adversarial with government reps monitoring every move, every step along the way. In the commercial mode, it was "we're all in this together, so let's come up with a good result."

Soon enough, just as the new ground rules got underway, the word came down from on high that, all other things being equal (each contractor meeting the six basic requirements) price would be the key discriminator, and it had better be well below the 40,000 dollar ceiling. About the same time, our missile division got an extra incentive to make JDAM work: they lost a major competition to be the sole-supplier of the Navy's Tomahawk missile—a program they had shared for the past ten years with Hughes, and for which they had produced more than fifteen hundred units. More layoffs were just around the bend and the word was, we lost on price.

JDAM Program Manager Charlie Dillow gave his team the charge: we have to cut costs. Ask questions, get answers, get affirmation, and make decisions. What's the life expectancy of a JDAM? Storage time measured in years, operational time in seconds. Parts must be made of stable materials, but there was no need for ten-thousand-cycle actuators. Which component is the priciest? What are the most expensive parts? Does that wire clip need to be made out of metal, or could it be plastic?

As noted earlier, we had established Integrated Product Teams (IPT) as a key part of the New Business Activity, bringing engineers, support specialists, financial managers, and so forth to join with marketers and salesmen, all under the coordination of business development. For JDAM, DoD picked up on, and added to, this idea, creating IPTs that combined government and contractor personnel. This was, I think, the first time the phrase in the running joke about the biggest government fib, "We're here to help you," actually meant what it said. DoD sent experts to function as if they were part of the manufacturer's own crew, to offer the perspective of both operators and procurement specialists.

We brought suppliers in as members of the team, where everyone—after signing confidentiality agreements—could freely share information, an engineer from one company could make a suggestion to an engineer from another. With enlightened self-interest, suppliers would offer solutions that reduced their own business but increased the chance of a win. One suggested if the antenna design was changed, they could drop some components from the circuit board. Someone asked, do we really need four moving tail fins or would three work as well, or even better—like the empennage on an airplane? Cut out a motor and controller, and to maintain stability, replace the fourth fin with a cheap, fixed unit.

The expanded teams turned out to be more successful than anyone might have expected. There were some awkward moments while everyone learned the new rules and there was, I think, some initial suspicion that competitive information would leak around the edges, but in the end, the IPTs really did work. The government folks and suppliers on our team were just as determined to win as we were. To win, that is, for McDonnell Douglas.

There were other departures from the norm, notably adding a progress "report card" that covered contractor performance at three points during the run of the EMD contract. There were color coded grades: blue for outstanding, green for meets expectations, yellow not quite there but may make it, red, bad news. DoD called it a "rolling down-select," where each team knew where they stood, and why, at each grading. Under the usual practice, DoD revealed nothing until the announcement of the award, and then it was more or less who won, who lost, and the losers would jump in and ask for a re-count, protest the award, hold everything up and cost everyone money. With the rolling down-select, there would be no unpleasant surprises at the end and there might be chances to bring up the grade in the interim. At McDonnell Douglas, we had an incentive-compensation (IC) plan tied to the grades to motivate people to focus on things deemed most important by the government. We turned the report cards over to a Win Strategy Steering Committee (WSSC) for analysis; the wise men (and women) might say, "Oh, you're having trouble with X. Have you thought about Y?"

The award was announced in the fall of 1995. Our winning bid was 14,000 dollars per unit. In the ten years since JDAM entered service, the price has drifted up to 22,000 dollars (half from inflation), but the planned 40,000 buy jumped to more than 200,000 units. That was a pretty good deal. Some of the DoD innovations and some of the facets of the "trial run" seem to have dropped off the screen on later competitions, but the color-coded interim report cards have proven to be of great value.

Harpoon

During the 1967 Six Days War, the Soviet "Styx" anti-ship missile captured headlines and the public imagination when, launched from Egyptian missile boats, four of them took out the Israeli destroyer *Eliat*. David versus Goliath? Perhaps. This action spurred the development of similar weapons in a number of countries. McDonnell Douglas, always ahead of the pack, had been funding studies for an anti-ship missile since 1965. After *Eliat*, the U. S. Navy came aboard, and the result, Harpoon, became operational in 1977. (As noted on page 8, Harpoon was my first "sale" with the company, when I convinced some members of Congress not to cut funding.) Not to be outdone, the Norwegians came up with infra-red guided Penguin (especially suitable for cold weather); the Swedes, RBS-15; and Aerospatiale, the French company, developed Exocet, which entered service in 1979.

Exocet was first to draw blood, which gave it a warlike reputation that it didn't really deserve. During the 1982 Argentine–U.K. war in the Falklands, two air-launched Argentinean Exocets were credited with sinking two British ships (including the destroyer, HMS *Sheffield*), and one fired from a truck damaged another ship. It looked great on the record, extolled by the French ever since in promoting Exocet. Well, it seems that two of the three did not explode, although they caused significant damage from kinetic energy (one punched through a rusty hull) and fire fed by the missile fuel. Exocet later played a big role in the Iraq-Iran tanker war of 1980–1988, seven hundred were fired and they hit a lot of ships. Nearly all of them were defenseless tankers and other merchantmen, but also—by accident?—the frigate, USS *Stark*, was struck by two, one of which failed to explode. More on this in a moment.

By 1989, the U.S. Navy (and Air Force) had more than three thousand Harpoons in inventory, the threat from the Soviet Union was going away, and that was that. The Navy stopped buying Harpoons. The future was in continuing international sales, which had begun with Turkey in 1974, followed by the U.K., Australia, Korea, Israel, and Japan, and then Singapore, Pakistan, Thailand, Egypt, Indonesia, and Greece through the 1980s. Sales had been brisk, thanks to the combination of a better product, a lower price (because of the large accompanying U.S. Navy buys every year), and FMS credit financing to many countries. But the future was looking a bit cloudy. Customers were musing, if the U.S. Navy wasn't going to buy any more, why should they? Will support be there when they needed it? We may assume that the folks from Exocet—Harpoon's only real competition—were happy to suggest an answer.

In 1991, Martin Fisher, who had just come aboard as director of Harpoon marketing, learned that Aerospatiale was about to sell Exocets to the Kuwait Navy, which already had some in inventory. Fisher wanted to offer an alternative. In Kuwait, someone handed Fisher a paper, written by the French, filled with unsubstantiated claims, half truths, and mistruths on the weaknesses of Harpoon. Just as our team did in countering the "Falcon Facts" junk in Korea, and despite the "company policy" that we never speak ill of a competitor's product, Fisher had to come up with an accurate rebuttal, overnight. He did, and managed to forestall the sale.

But Aerospatiale's aggressive tactics were bearing fruit in other markets and Fisher had to launch a broader attack, and here is where he got lucky. Someone mentioned the Tanker War, and later someone else said, "Those ships were all insured. Why not check with Lloyd's of London? See what's in the files." Check he did, and he found a treasure. There was data on two hundred Exocet launches against merchant ships, testimony of skippers, along the lines "I saw this missile fly over, just above the deck without hitting anything," and "I saw one splash by the side, a miss." There were also photos of unexploded warheads clearly marked with the Exocet model number. The known dud rate was about 70 percent. As for the non-duds, skippers discovered that water-filled balloons hung around the vulnerable bridge could trigger the warhead and absorb much of the blow. Fisher put together a killer briefing, "Harpoon Myths and Facts" supported by data from an unimpeachable source, Lloyd's of London. Great information to pack into the sales kit for the future. If there was going to be a future. Here was another program with dwindling production and a line threatened with shut-down.

The sales target for prior years had been one hundred missiles per year. In 1992, the company sold only fifty-six, a number circled in blazing red on the tote board. This was just after the LHX, A-12, and Korean losses. Here was some sort of handwriting on the wall—if the company was going to stay in the international missile business, they had to take action to do something to make Harpoon more attractive to wavering customers, and hence more competitive with Exocet, but, what? Harpoon hadn't been improved since 1983.

Ah . . . a thought. In the baseline model, if Harpoon missed the target for any reason, it would just keep going until it ran out of fuel. Why might it have missed? Because the target may have turned on some countermeasures. Okay, what about putting in a program that says to itself, "I should have hit the target by now. I wonder if it's really there?" and has the missile circle back for another look, and back again. Give it a larger fuel tank to accommodate this wandering around. Perhaps by then the countermeasures will have been turned off, and the missile will find and hit the target. It would cost about 6 million dollars to add this new feature to the mission computer and get it into production, but it would not be a complicated nor time-consuming effort. The company worked with Texas Instruments, which built the front end of the missile, and came up with a 50–50 cost sharing, 3 million dollars each.

Even though the Navy's Harpoon office had stopped buying the missile, they still controlled international sales through the Foreign Military Sales (FMS) program. The Navy agreed to support the change, give it official sanction, and to allow the companies to recover their investment through slightly higher pricing. The new version was designated Harpoon Block 1G.

International sales went from 56 in 1992, 93 the next year, then 89, 135, and then 209. We got the 6 million dollars back by the second year.

A good start, but Exocet was still pressing, playing catch up, and the market was demanding more. So, what might come next? The baseline Harpoon was fired toward a known target at a known direction and distance and at some point the radar would turn on, lock on, and guide the missile to impact. If some intruding element got in the way—for example, an oil platform—the missile wouldn't know the difference and would take out whatever came first. You couldn't use Harpoon against shipping parked in a harbor as there would be too much clutter returned from buildings on the shoreline. Even if the ships were moored well out from shore, the missile couldn't select one ship over another because they all look alike on radar. Ah, but add a GPS-aided inertial guidance system and you could lay down a mission plan to steer intricate paths to the target going around defenses, land or sea, and avoiding off-limits areas. Thus, the anti-ship missile became also a land attack missile. Feed in the GPS coordinates of the target, turn off the radar, and send Harpoon to a collision with a ship in port, a building, a shore battery, or whatever.

There was a counter-argument: why would anyone send a very expensive missile to do the job that a JDAM could do? Well, a Harpoon-equipped ship steaming offshore could attack a priority land target when no other assets were available. Also, very few maritime nations have JDAM-capable airplanes. They have ships, some have submarines, and—call it a cultural thing—most naval officers of whatever nationality want to be able to do the job whatever it is without having to call on the Air Force.

Where were we going to find a handy GPS unit and inertial navigation system? Why, in the building next door where they were assembling another low-production-rate missile, the Standoff Land Attack Missile (SLAM), for the Navy. Technically, it wasn't a problem, but there were some financial issues. Developing and testing this upgrade was going to cost between 60 and 80 million dollars. Serious money for a program that was small potatoes compared with the company's core business—fighter aircraft— and no one was going to approve that sort of outlay without seeing a solid business case. Fisher's sales team looked at the potential market ten years out, who might want to buy this seriously enhanced missile? (They already knew it wouldn't be the U.S. Navy, with anti-ship requirements covered by Harpoon and land-attack covered by SLAM.) The team came up with a rational estimate—all international.

However, as with the Block IG, the team needed approval from the Navy to offer the new model to the international Harpoon customer base and permission to tack on a surcharge to recoup the cost of development. The U.S. Navy gave the green light to develop, test, and sell, and to allow the company (by this time, Boeing) to recoup the investment.

But—hold on—the *Navy* said the company could sell it to the customer base, but when some members of Congress learned that we were going to offer a land-attack capability to certain countries, they said "No." So a couple of customers were dropped from the list.

Development of Harpoon II began in 1998, and it was delivered to the first customer, Denmark, as an upgrade in 2002. I must note, sales have been steady but not robust. Call it a lesson learned. In many industries, when you announce a new model, it's not hard to sell the old model, just offer big discounts to move out the inventory. In the missile business, we don't have an inventory and when a customer knows a new model will be coming along, he's usually willing to wait however long it will be until the new model is ready for delivery. There will be some vacant spaces in the order book for many, many months. It's something to think about when you are about to make the upgrade commitment.

Since the dismal year of 1992, the company hasn't lost many competitive sales but for a couple of those it did lose, the reasons are instructive.

- In a competition between Harpoon, Exocet, and the Swedish RBS-15, the Swedes won. Going in, we all assumed that this would be head-to-head between the first two, clearly superior to the third and both already in the customer's inventory. We didn't see any logical reason the customer would complicate logistics by adding a less-capable performer. There wasn't any *logical* reason. After listening to some public mumbling about a better industrial participation package, we were privately given another explanation: the decision maker had family ties in Sweden. Was this key? Who knows? Proba-

bly not the sort of thing you would think to explore when going through the "Decision maker and influencers" exercise.
- A Southeast Asian nation bought a couple of surplus Chinese frigates. These came with installed missile systems, which were quickly demounted and dumped. The competition for a replacement was between Harpoon, Exocet, and a different Chinese system. Our team had very good competitive intelligence on Exocet, knew pretty much everything they might put into their proposal and had great confidence in our offer. We drew a blank on the Chinese, but, since the customer already dumped a Chinese system, why would they buy another? Why, indeed? Because the Chinese also had a surplus submarine, which they tossed in to make the deal.

Four decades after entering service, Harpoon sails on, scrappy as ever. No warhead has ever failed to detonate in live firing. Ten missiles have been fired in combat against six targets; every missile hit and every target sank. For seventeen years, the only Harpoon sales were international and the only developments came from company (and supplier) funding. It's ironic, but as of today, the international community has significantly more capable and flexible Harpoons than the U.S. Navy. This is about to change. The Navy is buying upgrade kits to bring older models up to the standards of the latest version, Block III, which adds a data link for positive control throughout the flight.

The Harpoon program exemplifies the value of R&D programs in front of need, of solid and continuing relationships with your prime customer, and of a disciplined approach to every potential new development, every sale. Call it a mantra: "Is it real? Can we win? Where's the money? How long will it take? Can we meet the most important requirements?" If you can't answer those questions in the affirmative, you should find something else to do.

Malaysia

In 1993, things were really getting busy. We knew that Malaysia was ready to buy some airplanes, and also knew that they may already have made a decision to buy Russian. At least, such had been the thrust of an announcement in the local English language paper, the *New Straits Times*. Yes, we monitored newspapers.

John McDonnell and I flew over for an exploratory visit, and the American ambassador—who agreed with me that the uncontested selection of the Russians was a bit of an embarrassment—said, if we would commit to a campaign, he'd work on the Malaysian government to open the selection to a competition for all comers. He was as good as his word.

As a practical matter, "all comers" included MiG-29, F-18, and F-16. With the advice and assistance of a first-class consultant, a very-highly-regarded Malaysian banker who knew pretty much everyone who counted, we put on a full-court-press, involving our teammates Northrop and Hughes, subcontractor General Electric (engines)—GE CEO Jack Welch paid a visit to the prime minister—and with heavy assistance from the U.S. government. With the PM's science adviser, I co-chaired an economic impact study on the Malaysian aerospace industry. McDonnell Douglas sponsored a conference, "Pacific Dialogue," with Henry Kissinger as keynote speaker. As it happened, the prime minister was a former student of Dr. Kissinger.

To offset Russian tactics (gifts, questionable barter deals, rumors of promised "contributions" to the ruling political parties) we took the fight right to the folks who would be operating the airplanes. From our office in Germany, we obtained data on the reliability (poor) of former East German MiGs that had been brought into the Federal Republic's inventory. The ambassador arranged for the defense minister to be flown out to a U.S. Navy carrier to observe flight operations. In any event, Malaysia split the fighter buy, eight F-18s and eight MiG 29s, but they also bought thirty Harpoons and five MD-11s. That counts as a win in any book.

The chief salesman-consultant for the Russians was so impressed that, about four months later, he showed up at my office in St. Louis and asked, would I like to be his business partner for sales in the region? I thanked him for the kind offer, but said, my career was with McDonnell Douglas and I really didn't see myself as an independent gun runner in lower Asia.

Sometime later, perhaps more for geopolitics than national defense (the PM saw himself as the nexus for a new Pan-Asian union, for which he felt he needed Russian support), Malaysia purchased another twelve MiG 29s. As of this writing, reliability of the MiGs has been spotty and many need parts that are not readily available. The F-18s, on the other hand, endure, cornerstone of the Royal Malaysian Air Force.

Netherlands

Here's an example of political courage. We were negotiating for the sale of some Apaches to The Netherlands when our embassy got word that the prime minister had decided they would go with the Eurocopter, which, like the Apache, was just then coming on the market. Apparently the PM thought that if the Dutch were the first country to sign up for a buy, they might get a reward in the form of reduced pricing. Our man in Holland sat down with the U.S. ambassador and worked out what must be among the snottiest letters ever sent from an American ambassador to the prime minister of the country to which he was posted. The thrust, in essence: "You invited the United States into a competition for a new attack helicopter; you asked for pricing and availability, which we developed at a considerable expenditure of taxpayer financial resources, and I simply don't understand how you could make a selection without conducting a serious review of all candidates." As it turned out, the Royal Netherlands Air Force—the most interested party to this transaction—had not been consulted, and the ambassador's rebuke gave them some room to maneuver.

In the meantime, the company representing Eurocopter warned the prime minister, if they didn't get the order, the Dutch company Fokker would go bankrupt. Fokker had been in financial trouble for some time and apparently was in line to pick up some of Eurocopter manufacturing. The Dutch Air Force told the prime minister, in essence, "We're not saying that you have to buy the Apache, but we're saying, if you don't, you'd be better off buying nothing. There is no suitable second choice."

The prime minister opted for national defense over popularity. The Netherlands leased twelve AH-64A versions to begin training in 1996, and ordered thirty AH-64Ds—our first Longbow sale—the last of which was delivered in 2002. McDonnell Douglas tried to stave off Fokker bankruptcy with an infusion of 50 million dollars in C-17 work, but to no avail. The company soon went under.

Harrier

Here's an interesting partnership. The AV-8 Harrier, the plane that can take off, hover, and land like a helicopter or fly like an airplane, operate from a grass field or a highway, drop bombs, and get back quick for another load, was invented by the British. In 1970, the U.S. Marine Corps bought 110 of the first version, AV-8A. McDonnell Douglas was given a license to provide support in the United States. About five years later, our engineers came up with a scheme to increase range and payload and developed the improved AV-8B Harrier II for use by the Marines and the RAF. Deliveries began in 1984.

The British contractor, BAE, began searching for international customers. Spain was on their radar, but diplomatic relations between the two countries had broken down over some perennial dispute over Gibraltar. So they asked if we could help out, take over as prime contractor for the sale and be 50/50 on production. We made the sale to Spain and reached an agreement with our partner to divide up the world for sales of the land version.

BAE kept maritime Sea Harrier sales for themselves, which soon enough led to some conflicts. We were trying to make a sale in Italy when we discovered that BAE was at the same time hustling the Italian Navy. It made for a bit of tension, but we got the sale. Five years later we set up a co-development agreement with both Spain and Italy for the next advanced version, the radar-equipped AV-8B Harrier II Plus. This turned out to be a very successful program.

If this starts to get a bit confusing, bear with me. At this point, the Marines had a dream: replace all seventy-four of their old day-attack versions with the Plus, but they didn't have the money and couldn't get the approvals. This is where I got into the act, but not for the better. I threw our AV-8 salesman out of my office because he had a crazy idea to sell the Marine Corps on a program to take the aging Harriers and bring them up to the Plus standard. "We sell new airplanes," I yelled. "Goddamn it, not repainted dinosaurs. The Marines want the best. Go find something useful to do!"

What I should have done was to follow my own teachings and ask a few questions. Does the customer have a real need? Is this based on credible data? Can this be translated into a requirement document and funded? Can partnership opportunities increase the chance of a sale? Can we line up sufficient support in DoD and the Congress? Can we deliver? The answers all would have been "Yes."

Well, our AV-8 salesman was a recently-retired Marine and knew his way around the system. He worked out a deal where we would fully remanufacture all the day-attack aircraft. He brought DoD and Congress aboard by showing the combat advantage of the Plus and the cost-advantage of remanufacture over new production.

As for partnerships—he heard that the Naval Air Depot at Cherry Point, North Carolina, which handled Marine Corps aircraft, wondered why they couldn't do the whole job? He easily convinced them that there was more involved that it might seem, but that they could play a major role, picking apart the old airplanes and upgrading or refurbishing those parts that would be re-used. Our British partner produced a new aft fuselage structure, Rolls Royce contributed new engines, and McDonnell Douglas handled overall coordination, systems engineering, production of other new components, assembly, test, and delivery.

In addition to airplanes for the Marines, we remanufactured five more for the Spanish for a total of seventy-nine, with deliveries between 1996 and 2003. The remanufactured airplane has 100 percent of the capability of a new Harrier II Plus at two-thirds the cost and full service life of 6,000 flight hours. The program resulted in more than 2 billion dollars in sales.

Thailand

Here's another example of a done deal turned sour. In 1987, Thailand purchased some de-tuned (by U.S. government fiat) F-16s and took delivery in 1988. Around 1998, ready to move up, Thailand posted a competition between a less-restricted F-16, the un-restricted Russian Sukhoi Su-27/35, and the F-18C/D. After a spirited tussle, they selected the F-18, with an order for eight.

Good news? Not really. For reasons I cannot rationally explain, they cancelled the order before delivery; something about an economic crisis in Southeast Asia. Yes, there was some sort of general crisis, but I could not then (and won't speculate now) as to why canceling the purchase of eight airplanes would save the day. Especially since, within a year, they put in an order for sixteen refurbished and less-capable surplus USAF early model F-16s. I do believe it had something to do with a change of government and the new guy didn't want to validate anything the old guy had done. I also believe this was another case where we lost an important sale because our people lost touch with the customer.

F-18 "Upgrade"

About the same time we were working with the Swiss and Finns, an F-18 marketer made the rounds of Marine Corps attack pilots, looking for some business to develop. The Marines flew the Grumman A-6 Intruder, a carrier-based medium attack plane that first entered service more than twenty years earlier. "What do you need?" he asked. Oh, this and that, more modern equipment, greater range. "What about the F-18?" It was okay, but not good enough in the attack role because of limited load and limited range.

Our man came up with a plan for an improved attack version of the F-18. He took his idea to an executive committee. They listened and said, "Very interesting," but offered no real encouragement. He worked out some details, and when he was at the point where he needed funding to continue, went back to the executives, who said, "Very interesting," but did not approve any money. He gave up in disgust and went on to something else.

What he didn't know and the executives couldn't tell him was the company was teamed with General Dynamics on development of a top-secret carrier-based stealth attack plane, specifically intended as a replacement for the A-6 and several others. This was the A-12 Avenger II, the plane later cancelled by DoD.

But by 1991, with A-12 off the table, there were no projects to replace the aging fleets of Grumman A-6, EA-6, and F-14; starting development from scratch would take a very long time. Someone had the bright idea to upgrade the F-18, but to meet the varied mission requirements, we would have to do more than just bolt on some new equipment. The plane needed a larger airframe, stealth capability, more power, a lot more fuel—virtually a whole new airplane. But at that moment, Congress was not in the mood to fund new airplanes and we didn't want to get Grumman all excited. It was time for a work-around. With the collusion of several government agencies and some mid-level Navy officers at the Pentagon, the company worked it up as an "engineering change" proposal, a category for which funding was available. It was probably the largest ECP in history. By the time anyone discovered the plan, it was too late. The F/A-18E/F Super Hornet entered service in 1999.

Military Aerospace Support

In 1995, Jim Restelli was named head of the recently-titled Military Aerospace Support Division, and brought along a vision. Here, we had a company that built airplanes and provided after-sale product support for some of those airplanes. Outside the family, there was a universe of military aircraft, of many parents, that needed support. Was there an opportunity? I remember well Restelli's pitch to the management of the newly-merged McDonnell Douglas and Boeing: the design, development and production of military aircraft make up only 30 percent of the government's investment in total ownership. The bulk of costs went for program planning and management, training, technical manuals and support equipment, maintenance and modifications, upgrades, and a heck of a lot of aging aircraft initiatives. You could build a multi-million dollar business just by tapping into a chunk of that 70 percent, logically focused on modifications, upgrades, and modernization.

Restelli wanted the company to make a substantial investment, to add or re-purpose facilities, and was looking for approval. His timing was excellent, as a Base Realignment and Closure Commission (BRAC) had just called for the closure of a number of military installations, and chambers of commerce nationwide were begging companies to come take a look, to take advantage of recently-refurbished facilities and a trained blue and white collar professional labor force about to be cut adrift. One of those locations was Kelly Air Force Base, San Antonio, which operated a large overhaul depot.

And, at about the same time, DoD wanted to offload some functions that could easily be handled by private industry, including logistic support, and was encouraging "alliances" between industry and the military.

One of the first opportunities: there was to be a competition for civilian depot maintenance of the KC-135—an airplane first delivered in 1965 and likely to be in the fleet until at least 2040. Yes, it was a Boeing-built airplane, perhaps an advantage, but there were a lot of full-time maintenance and overhaul companies, many of which handled major overhauls for commercial airlines, who would be lined up for the opportunity. Clearly, a DoD-encouraged "alliance" with the right government partner would be a major discriminator, and Restelli lost no time negotiating a memorandum of understanding with the commander of the Air Logistics Center at Ogden, Utah.

The award went to Boeing, in a contract providing a 50/50 split in KC-135 depot work between the company's new facilities in San Antonio and the Air Force Depot at Tinker AFB, Oklahoma City, Oklahoma. This was the first step toward building what now is a six billion-dollar-a-year business, and which is still growing.

Here's an example of another opportunity. The Air Force had 519 C-130 workhorse airlifters in service, some dating from 1962. They were excellent airplanes but with rapidly-aging avionics, radar, flight management systems, and navigation systems. There was no "standard" C-130 cockpit; the force included fourteen variants of five different models, created for a variety of special purposes including tanker, airborne hospital, weather reconnaissance, special operations, and gunship. And none of the aircraft could connect to the growing global air traffic management communication system.

It was time for upgrades and standardization, under a formal Avionics Modernization Program (AMP). There were four proposed solutions for the C-130 AMP, four bidders for the job: Lockheed Martin (the original builder), L-3 Communications, BAE, and Boeing. This was the first time Boeing had offered to take responsibility for an aircraft not originally produced by either McDonnell Douglas or Boeing.

Was there opportunity? Certainly. What did Boeing bring to the table? Beyond a solid reputation and status as the world's largest aerospace company, Boeing brought experience with modern avionics and cockpit design gained in the C-17 and commercial aircraft programs. They also brought a positive track record working with the Air Force Special Operations Command on C-130 gunships and successful demonstration of break-through state-of-the-art non-proprietary "open" military computer systems. These were created by the McDonnell Douglas "Bold Stroke" project—systems that cost less to develop and produce than then-current "closed" systems, that are easier to maintain and upgrade, and had been proven in the F-15, the F-18, and the AV-8B.

Opportunity, yes, and confidence, but why make the investment to compete against such formidable opponents? Beyond a possible 4 billion dollar new revenue, the experience could transfer to development of a next-generation tanker. The chance of a win? Pretty good. And win they did.

CHAPTER 5

The Process

Well, dear reader, we—you and I—have just completed a metaphorical journey through the twists and turns of the aerospace business, a short course on assembling and deploying a sales team, avoiding trips and traps, and vanquishing the competition. You've watched as McDonnell Douglas marketing moved from organizational chaos with a reliance on tribal knowledge, through a tentative approach to order and discipline, to the development of the New Business Activity and the 12 Steps (for a refresher, re-read pages 30–35)—and began winning more sales.

We've now arrived at the grand finale, the reason for this book, a summation of all the wisdom and advice that have been scattered along the trail. And more. Our efforts to get better, to do better, didn't stop with the 12 Steps, cast in stone, but evolved into what I've called a full-fledged, proper name with a capital-P—Process.

The Process became the basis for an expanded marketing program at the merged McDonnell Douglas-Boeing. Other companies have applied similar approaches in their own businesses, some derived from our effort, but I'll tell you how my team and I did things; you can mull this over and decide for yourself if what we did is a good fit for *your* business or if you should make adjustments. As presented here—our version, developed by many superb teammates and mentors—the Process is flexible, easily adapted to explore new markets or to maintain and grow existing business, and to develop successful campaigns to win in either.

In the most significant improvement to the 12 Step[s], checkpoints—gates—guarded by suspicious executi[ves] or withhold, funding for the next leg of the journey. Be[fore] the marketing team leader, the program manager, and must answer questions posed by the Process: have you d[e] considered the probable moves of the competition, the resources that would be occupied, and posited the engineering challenges that must be met before moving ahead? For the first gate, you calculate the probability of winning and decide, do you even want to get into this fight? If you can demonstrate that the probability of a win is high and, yes, you want to enter the lists, you get approval to advance to the second gate. There, you present your strategy for winning. Look good to the executives? Move ahead to the third gate, where you have assembled all relevant information and present the company with a decision: to bid—or not. At the fourth and last gate, your proposal is reviewed and the company must make a decision to submit, and be in the fight or not.

It was, at first, hard to convince program managers of the value of the gates (officially, called Executive Reviews) in large part because the program managers didn't want to expose defects in their approach (or in their intelligence) to the executives. It was even harder with some of the small program leaders, especially the pet-rock keepers running hobby shops in pursuit of research topics that might or might not have any relevance to the plans or prospects of the company. A typical reaction was, "Oh, this is not a very large program, we don't need to take up a lot of your time with a proposal or bother the executives. I'm sure they have more important things to deal with." Wrong.

People quickly discovered that it could be quite embarrassing to come up against a gate and not have an answer for the executives assembled. It could be career limiting, but at the least delayed progress until the responsible program official could assemble enough solid data to push through the gate. It may have taken only two such embarrassments before the word spread, "Be prepared."

Perfect adherence to the Process will not guarantee a successful outcome. In fact, it could be the path to failure, unless supported by critical thinking, good judgment, and hard work. The team makes the sale; the is only a tool, a disciplined approach to organizing complex activities. It frontloads the collection and analysis of all sorts of data, some of which will require further investigation and study before you're ready for the first gate. Think of it as a roadmap, with checkpoints along the way. Follow the map, answer the questions. "Who will make the decision to buy the product?" Don't have an answer? Get one. "Who are the people who will influence the decision-maker?" "Who on the team will be responsible for working with the influencers?" Probe. Question. Take nothing at face value. Someone may say, "The Navy wants it to have a swing wing." "By 'Navy,' you mean the chief of naval operations?" "Well, no not really, it's the commander Naval Air Systems Command, NAVAIR." "Is it really the commander?" "Well, no, not exactly. . . ." And there must be someone tracking progress, checking off the items on the list one by one until all are done.

The "gates" are set up as, well, gates, to actively involve senior management: you can't go from one to the next until you have permission to pass through. But overall, the Process really begins well before Gate One and continues long after Gate Four.

And so, without further ado, let me introduce the full-up, full-scale, fully realized Process, in a thorough, step-by-step exposition, incorporating much of what you have already read and much that is new, here presented as eight discrete phases.

Phase 1
Identify and prioritize opportunities, decide what you want to do next (pursue one or several opportunities), tell management, and get approval at Gate One.

Are there new technologies you ought to capitalize, anything coming out of your company research center? What's your competition been doing lately? Are your current customers planning any new initiatives? Can you spot any potential new customers?

Unless you already work for a large company with a broad and varied product line, it may be hard to grasp the range and depth of opportunities that will come your way almost every day. How, indeed, do you decide which to pursue, and which not? First, you must have a core vision that defines the kind of business you want to pursue and explains where you want to be in ten, twenty years. Perhaps you want to be the leader in "air superiority," as good a place to start as any. Next, for solid political reasons, you probably want to have a stake in each of the key customers—the active Air Force, Navy, Marine Corps, Army, with some dealings with each service reserve force. However, not all of these buy air superiority, so your list of competencies must expand to cover such missions as anti-ship, anti-personnel missions, and transport. You might also want to include support and R&D.

Put simply, know where you want to go and how you plan to get there—have a framework within which to assess opportunities. I'm not just talking about adding new products, but also mergers and acquisitions, another fertile field. Too fertile, actually, as too many companies are eager to sell themselves to you, offering new capabilities to enhance your products and services. Joining forces with some may be worthwhile while others may be better brought in as contract suppliers. Sort it out with a disciplined process.

Next, let's say, you've been told that Country C is ready to buy some airplanes. Is this a real opportunity, or is this just wishful thinking on the part of Country C, known for having a fragmented, contentious government and a thin bank account? Perhaps there have been some changes: a new government, or they struck oil. Check it out. Perhaps this customer always asks for more planes than they can buy. It's not because of lack of funds but because they don't have the support facilities ready or can't train the pilots and maintainers fast enough and can absorb just so many planes at a time. Check it out.

Why does Country C want to buy this product at this time (replace aging fleet, counter neighbor's arms build-up along the border)? For that matter, what do you know about, or can you learn about, Country C? Where do they want to be ten years from now? What is their acquisition process? Do you know anything about their support capabilities? That is, are you about to try to sell a very complex machine to a country with one foot still in the middle ages? Will the mechanics be able to keep this airplane flying? If not, what could you offer by way of in-country support? What are the Most Important Requirements (MIRS)?

Don't ask the customer directly, what his values are. You'll get the school-approved answer and learn nothing. When you ask a military guy, "What do you see as the prime driver here?" and get the answer, "Price," don't believe it for a moment. There's a world of difference between selling commercial aircraft where price is everything and selling in an environment where the guys who are buying your airplanes don't have to pay for them—it is not their money! Military guys want performance and most of them don't much care what it will cost. That's for Parliament or Congress to worry about. Or maybe the guy says, "value." Maybe. You have to pierce the fog. I don't think we ever nailed what the customer really wanted—the customer, that is, the admiral or general above the guy you're talking to, or his boss.

One would hope that you already have been cultivating Country C, perhaps through a local office established after an earlier sale, or by regular informal contacts. In any event, now is the time to spread out across the globe and start talking with people, not just key people in Country C but anyone who has worked with or understands them. Ask—and listen. As my third grade teacher, Sister Mary Charity, said: "God gave you two ears and one mouth. A design to encourage that you listen twice as much as you speak!" Wise counsel.

While your "customer" may be identified as the Country C Air Force, that's just a label on a collection of players within, connected to, or superior to the Air Force. You need to know, and work with, all. You might start with the users, the pilots, maintainers, commanders. What are they looking for? Then search out those players above the users, who must juggle myriad special interests, balance budget issues, who may offer a different perspective. Next, the politicians—are they interested in hardware or prestige? Defense or muscle-flexing? Don't forget the evaluators, who need to know about your company long before they read about it in the proposal.

You should also understand that in the United States, at least, the evaluators don't usually pick the winner. They forward recommendations, "These bidders are all in compliance," separated by a few points up or down in various areas. Then they wait for the body language, which they get in a preliminary review with the source selection board, and perhaps issue a final draft reflecting just what the source selection board wanted to see. Now, suppose the head of the board is the secretary of the Air Force. You can bet that he has already had informal, direct, non-discoverable guidance from the secretary of defense. Sometimes the apparent decision maker may just be ratifying a decision made further up the line. I've asked my staff, "Who have you been working? What influencers?" and they usually give me some names, all from the lower end of the pyramid. I don't think anyone ever said, "secretary of defense."

So, how would you work the secretary of defense? You might encourage senior members of the Senate Armed Services or Appropriations Committees to request personal briefings on the status of a program, which just happens to be of particular importance to them.

Don't send just anyone, no matter how good a listener, as a "customer contact." These are critical assignments, which can do a lot of good for your campaign, or a lot of harm. A customer contact person must be open and affable, should have some skill with the local language, must know the product inside and out, know the messages to be delivered, and the appropriate timeframe in which to do so. It's sort of like choreography, where the steps must come in a certain order or the dance falls apart. Create a simple contact plan, listing who is to be contacted, when, by whom, and to what purpose.

Develop a "threat matrix" to plot the military posture of your customer's immediate neighbors and more distant sworn enemies. Where is each threat? How big is it? Is it ready now or growing? Has there been any public demonstration of recent antipathy? This will give you a good idea of how your product might be used by the customer, and when, and help you to show the customer how your product will be of maximum benefit. If this customer is an antagonist of another customer, but both are U.S. allies (the 1990-era Saudis and Israelis come to mind; see page 130)—what is the U.S. government's posture on arms sales to either?

Pull out the "lessons learned" file and review. Look especially for things applicable to this competition. Have a conversation with the key players in earlier competitions; they may offer some insights not in the file. Common errors at this stage are not putting enough people on the job, not staying close enough to the customer, and letting your competitors—who may have been staying very close to the customer, indeed—

influence what goes in the RFP. (See page 98) For that matter, what do you really know about your likely competition? Their market share, political connections, and strategic alliances? Have you come up against them before, and if so, to what effect? What's the word on the street—strengths, weaknesses, viability of the product? How do you think they will shape their offer?

Okay, so you have determined that the opportunity is solid. The next few questions: do you have a reasonable expectation of winning? Would you make enough money? Are there any reasons you might not want to bid, even if you feel you can win? For example, at any given moment there may be fifty competitions underway, large and small, and your resources are finite. Your engineers can handle just so many programs at a time. You can't afford to have your front-line sales people tied up in small sales when you need to get some big programs on the books at the same time. However, don't use the size of the sale as the sole determinant. Some customers might be considered so key that you'll pursue a sale at the expense of a sale to someone else. Israel used our products in what amounted to real-world demonstrations, with 100 percent combat victories. It may be a small sale to Israel, but a great selling point to other customers.

With the preliminaries over, it's time to get organized. Pick the first key members of the team: a leader and a finance manager. Go through Checklist A (below), present your case to management, and get funded. (At McDonnell Douglas, from this point forward, funding was supported by the New Business Fund (NBF), created just for this sort of activity.)

Checklist for Phase 1

Exploring Selected Opportunities
- ☐ Is this a "real" opportunity?
- ☐ Does the opportunity fit with company core vision?
- ☐ For a purchase, ask "Why does the customer want to buy?"
- ☐ Develop "threat matrix."
- ☐ List Most Important Requirements (MIRS) and define the customers long range vision.

- [ ] What is their acquisition process?
- [ ] Do they have funding?
- [ ] What are their support capabilities?

Getting Organized
- [ ] Select a team leader and a finance manager.
- [ ] Create a contact plan: who is to be contacted, when, by whom, and to what purpose.
- [ ] Review the "lessons learned" file for this customer, and for similar opportunities.
- [ ] What plan would you offer?
- [ ] Are there export license issues?
- [ ] Are there any industrial participation issues?
- [ ] Assess the likely competition: strengths, weaknesses, and viability of their product.
- [ ] How do you think your competitors will shape their offers?
- [ ] Do you have a reasonable expectation of winning?
- [ ] Would you make enough money?
- [ ] Are there any reasons you might not want to bid, even if you feel you can win?
- [ ] Are you ready to move ahead? If the Executive Review says "yes," move ahead.

Phase 2
Develop your winning strategy, and get approval at Gate Two.

What will be your strategy for winning, and what actions must you take along the way? Now is the time to more fully staff your program and develop your plan. Many companies today call it a "Capture Plan," as in, capture the contract, and assign a "Capture Team Leader." I used to call it "Marketing Plan" and "Marketing Team Leader," but you can use any labels you wish.

Your team is not just a bunch of sales people (and one finance expert) but an Integrated Product Team (IPT); a multidisciplinary group including engineers, logistics, and manufacturing experts, people from the D.C. office and the field offices. It would be wise to also include a few lawyers or other specialists to deal with issues involving intellectual property (such as patents, trademarks, copyrights, and especially, source codes, always coveted by the buyer), export licenses (the U.S. government might prevent you from releasing those hot source codes), industrial participation (to refresh your memory, see page 102), or environmental considerations: different countries have different laws, regulations, and practices. And now is the time to bring in a proposal manager.

It's also the time to create your winning strategy. In essence, you pose the question, "What will it take to win?" and provide the answers: good price; support from the user community; solid offset proposal.

You're not finished, of course, with customer contacts and gathering information. You're dealing with a moving, ever changing world. Your competitors will have been going through a similar exercise. Where do you think they stand at this point? What are they working on now? It might be time to bring in the Red Team—a group of people with a good grasp on how your competitors and customers think and act—to review your efforts, tear apart your strategy, make fun of your assumptions, and provide a best-guess as to the competitors' offerings.

Balance all known factors, especially those involving your competitors, and create a preliminary draft of your offer. In this, you will answer all of the questions that senior management might pose and cover all of the points that your customer will expect to see in your written proposal, including:

- a clear description of the product (or service) you are selling, including subsystems, hardware and software;
- plans for support of the product in the field, including spares, training, any warrantees or guarantees;
- how you plan to organize and execute the program; possible teaming and alliances;
- your process for selecting suppliers;
- export licenses, intellectual property, industrial participation;
- and the customer's budget, how much you expect to charge, and how it compares with similar sales.

Don't stamp this preliminary draft as "Hold Close." You may strip out an odd paragraph or two for company-eyes-only (like, what you expect to charge), but, properly executed (and marked "Proprietary") the draft becomes part of your customer contact activities. It gives the customer a sense of where you're headed and lets you get some useful feedback.

Next, set up a war room where all members of the team can review progress, learn the latest developments, new requirements, and issues to be resolved by walking through the door and reading everything on the walls. Post campaign storyboards that outline every step to be taken and by whom. Include all elements for the proposal, a rough cut on all graphics, a check list to ensure that all proposal requirements will be met. Outline your plan to mitigate risk. (Will that new technology work as advertised? Bring in a consultant, a professional risk analyst, and put his comments in a footnote to add credibility to your own assessments.)

Changes to any display should be marked by colored flags for instant reference. The ultimate war room accessory for the ultimately big program would be a full-scale mock-up of the airplane, developed well before it is time to write the Proposal. If you are working far enough ahead—as you would on the ultimately big program—the mock-up can do double-duty as a whip-up-excitement exhibit at the big air shows.

Convene the Win Strategy Steering Committee (WSSC). The goal, of course, is to have the WSSC say, "Good job, keep it up." It isn't always that simple, but is always of great value. Finally, adjust your plans, and prepare for the Gate Two Executive Review, Get approval to move ahead.

Checklist for Phase 2
- ☐ Create a marketing plan and appoint a marketing team leader.
- ☐ Set up an Integrated Product Team (IPT) and appoint a proposal manager.
- ☐ Set up a war room.
- ☐ Update your file on competitors.
- ☐ Draft a winning strategy. Ask yourself, "What will it take to win?" and provide the answers.
- ☐ Crate a preliminary draft of your offer. What you are selling; how you will organize and execute the program; teaming and alliances;

suppliers; support; export licenses, intellectual property, industrial participation, and price.
- ☐ Bring in the Red Team to review your strategy and efforts to date.
- ☐ Share basic details of your offer with the customer, and get feedback.
- ☐ Convene a Win Strategy Steering Committee (WSSC) to analyze your efforts and assess your progress.
- ☐ Adjust your plans based on feedback from the WSSC and prepare briefing for Executive Review.
- ☐ Get approval to move ahead.

Phase 3
Design your approach, your campaign, and expand the team. At Gate Three, get approval to make a bid.
Now you should be off to a good start, but you don't have the finish line in sight. It's time to work up a schedule, although you'll have to estimate most of it. When will the RFP be issued? What deadlines will be imposed? How long might it take to write the proposal? This will give you some sense of staffing and other asset requirements.

Refine and update your offer, your campaign plan, and your war room displays. Review all of your assumptions, re-check your data, incorporate suggestions from the Red Team and guidance from the WSSC, and, especially, feedback from the customer's review of your preliminary draft. Designate a program manager (unless you're working with an on-going program, or the marketing team leader will be doing double duty). Expand your team with program folk and representatives of your major suppliers.

Throughout the Process, start early, you should have been working on "price." As you will appreciate, it's complicated for a new product, considering all of the elements that must be considered. That's why you have manufacturing, program, and finance people on your team. If your product is already in production, you have a pretty good idea of the cost to manufacture and support. In any case, you must understand the customer's position on price. Are they looking for the best product and they are willing to pay for it, or are they a best value or lowest price customer? How do they feel about risk? To ease the customer's troubled mind, you can offer specific warrantees and guarantees for the new navigation system or the redesigned missile launcher. If they don't work as promised, they will be fixed or replaced at no cost to the customer.

Lurking in the bushes will be your competition—maybe five or six companies who want this deal and will go to the mat to get it. By now, based on past-performance and current intelligence, you should have a pretty good idea of the position of each. Do they come in with attractively low prices but exclude certain systems or functions? ("Navigation system? That's extra.") Do they typically make deals, accept barter, or arrange financing? What sweetener can you offer? (We took old airplanes in trade, for example, and sold them off to buyers with less-sophisticated requirements.) If you know your price is going to be high, where can you trim? Delete a nice-to-have but not required feature? Drop an advanced material and put in something tried-and-true and cheaper? That's why you have smart people on your team, on the WSSC, and in senior management. Decisions must be made.

Are you ready to enter the competition for real? It's time for the Gate Three briefing to management and another approval.

Checklist for Phase 3
- ☐ Estimate the schedule: When will the RFP be issued? What deadlines will be imposed? How long might it take to write the proposal?
- ☐ Review all of your assumptions, re-check your data, incorporate suggestions from the Red Team and guidance from the WSSC, and, especially, feedback from the customer's review of your preliminary draft.
- ☐ Designate a program manager (unless you're working with an ongoing program, or the Marketing Team Leader will be doing double duty).
- ☐ Expand your team with people from the program and representatives of your major suppliers.
- ☐ Refine and update your campaign plan, your offer, and your war room displays.

- ☐ Analyze the customer's position on price. Are they looking for the best product and are they willing to pay for it, or are they a best value or lowest price customer? How do they feel about risk?
- ☐ Develop specific warrantees and guarantees, as appropriate.
- ☐ Update your assumptions on the competition, marketing strategies.
- ☐ Review pricing. If you had to, where would you trim?
- ☐ Move to the Executive Level briefing for another approval.

Phase 4
Develop a complete and accurate proposal and determine a winning price. Get Gate Four approval to submit a proposal.

The customer will issue the Request for Proposals (RFP) or an international equivalent; in truth, they may actually issue two versions: a draft, and sometime later, after errors have been eliminated and bidder's questions have been factored in, the final. In a typical timeline, you may have one month between receipt of the draft RFP and release of the final. You need to cram in as much customer time as possible during that period, to pose questions, get answers, noodle possibilities. Once the final RFP was in hand, in my day at least, your option to initiate communication ended. (The customer, on the other hand, will likely continue to have questions for you.)

The team leader and key personnel must read both the draft and final versions, every page, and not just the sections that apply to their specific areas of interest. RFP's have been known to hide important requirements in illogical spots. Need clarification on any points? Have any questions on anything? You might send them to the customer. However, now that the game is afoot, all of the questions from all interested parties (and all of the answers) will be posted for all to see. You may therefore be reluctant to ask anything because that might tip your hand to your competitors.

You already have most of what you need (reviewed, revised, and revised yet again). Bring in the writers, graphic artists, IT professionals—people who know how to do this stuff, know all of the rules and regulations and customer-imposed restraints—and have a meeting with the entire team, to let everyone know what everyone else will be doing. Establish clear guidelines: don't just brag about the features ("New compact radar, fits in half the space," "New lighter weight brake discs"), but address functions ("Detects enemy aircraft at 200 miles using half the power," "Carbon-carbon brakes will stop the airplane in half the distance of steel brakes"). Here's where you slip in some subtle arguments about why you should win along with some not-so-subtle digs at your opponent. Your airplane has phased array radar, the other guy does not. "We have chosen phased array radar because the alternative, all aspect radar, carries such a large technological risk."

Assign a team leader for each section of the proposal, who will keep track of every element in that section, make sure that all important points are included, and that all fit into a prescribed length. Also, assign someone to keep track of the keeping track.

Update the war room story boards and displays, make copies, pass them to senior management, engineering and finance VPs, manufacturing and advanced technology leads for an adult-level review, then hand them out to the folks responsible for one section or another of the proposal. This gives them raw material for their endeavors.

Set everyone to work, and watch the calendar. You may have a week, you may have two months, but you won't be allowed a day—an hour—longer than specified. For a large proposal, this will be a very big effort indeed, resulting in perhaps 100 books, each between 100 and 200 pages long. You may need to have three people working on each book. The graphics shop will be working two 12-hour shifts; the writers and editors 12 hour days, seven days a week with time off for good behavior, but only when the job is done.

Check and double check that you are meeting all the requirements: technical, management, finance, your record of past performance, and so on. As I noted on page 97: the RFP spells out exactly what must be included and in what order; leave something out and you will be considered "non-responsive." You must follow the specifications for page size and format, type font and size, with all set to standard default values. No squeezing in a few hundred words here or there by varying the line height or letter spacing. Why not? Call it the level playing field factor. You will be limited to the number of pages in each section. Your incentive to conform? "If exceeded, the excess pages will not be read or considered in the evaluation of the proposal."

Since your customer already has seen most of what is going into your proposal—at least, the important highlights—you don't want to hand them any surprises. Don't play bait-and-switch. You had talked about X units at a price of Y, got them really interested, but the official price is 2Y? Not a good move, even if 2Y is a valid price.

The words will be important, but graphics will be vital. Remember: your proposal will be studied by a whole bunch of people, some of whom will be rushing ahead to find the juicy parts. Good graphics will be your best friend, quickly conveying a message, a concept, and the benefits of your offering. I used the "20-second" rule. If the viewer doesn't get your intent or understand the message within 20 seconds, you lose. Good graphics will be even more important for an international sale. While English may be considered the "official" language of the aerospace industry, do you really believe that all of the evaluators, operators, and influencers will understand everything you write? For that matter, if not prohibited by some naive rule in the solicitation, you would be wise to provide translations of everything you provide even if they are not requested.

Language issues aside, never underestimate the sophistication of the customer, the evaluators, and the decision maker. In many ways, many of our international customers long ago set the gold standard while the United States lagged behind, our system bogged down in a combination of misplaced priorities and politics. Military careers were enhanced by duty with the combat arms, not in procurement. But today, more and more military officers are trained in the intricacies of contracting and acquisition. They are bright, aggressive, want the best for their service and they will track your every move, parse every word, and let you know where you come up short. (I can't speak to "politics," except to say that is an arena where I doubt if much of anything has changed.)

Have at least one "outsider" team run a progress review on the proposal (you'll need more teams for a large competition), and ask the reviewers to simulate evaluators. Do they get it, or would they want to ask some questions? You want everyone to get it the first time through. Yes, the customer may ask questions about your proposal, but if they have to do so, you're losing points, psychologically, if not formally.

Checklist for Phase 4

- ☐ Study the RFP. If there is a preliminary draft, look for ambiguities, possible problems, and post questions to the customer.
- ☐ Assign a team leader for each section of the proposal and someone to keep track of the keeping track.
- ☐ Bring writers, graphic artists, IT professionals onboard. Give them something to do.
- ☐ Constantly update war room storyboards and displays.
- ☐ Create and post the work schedule, keyed to RFP requirements.
- ☐ Keep a running check on requirements for content and length.
- ☐ Ensure that all elements are consistent with what you've already told the customer—no "bait and switch." If something has changed, have a very good reason why.
- ☐ Subject all graphics to the "20-second" rule. Adjust or re-do as necessary.
- ☐ Have the Red Team review the proposal and incorporate comments.
- ☐ Convene a final WSSC.
- ☐ Proposal ready? Cost/price scrubbed to the smallest detail? All comments incorporated, reviews complete? Get permission to submit, and submit.

Phase 5

Submit. Wait for the request for your Best and Final Offer (BAFO) and submit your revised, best and final, offer.

At this point, you can, and you most surely will, reduce the size of your team, but you must keep key positions filled and active with other participants on call as needed.

A most important caution: when the proposal is finished, printed, and bound, but not yet delivered to the customer, you must invoke a firm "hands off" rule. Put it in a big safe and keep it locked. You will find that senior executives, particularly some who may not have been paying much attention during development, will now want to see the finished product and of course add some value. It is too late. Convincing them of the wisdom of that will be difficult.

Once the proposal has been delivered to the customer, you may rest assured the customer will have some questions for you. Look for trends. Something in one section may be questioned and of course you will respond, but if the same question shows up across several sections, take warning. Something may be very wrong with the proposal and you need to figure out what. The questions will not all be combative or probing. In some, the customer may be trying to send you a message ("Are you sure you want to have an open cockpit?") or subtly give you a chance to rebut something a competitor may have "suggested" was a weakness of your product (as in, "Is it true that your combat radius is limited to 500 miles?"). Answer every question in a positive, can-do tone. Do not get into an argument with the customer.

On many, especially very large competitions, there are follow-up solicitations, each updating some aspects of those previous, as the customer learns more about what the competitors are offering and what changes, if any, might be advantageous to the customer. This back and forth will take months. Have patience and stay alert but don't bug the customer. In most competitions, contact will be limited to normal business that is only outside the scope of this effort.

You may soon enough have the opportunity to make an oral presentation. You most likely will be given a team and time limit, say, four people for not more than four hours. The customer gets to meet four of your people. (Make sure they are the "right" four—personable, relaxed, knowledgeable. Especially, knowledgeable, because they may be asked questions.) Rehearse, of course, and have a session with what the defense industry calls a "murder board" to pose the toughest questions anyone can think of.

Eventually, you will be asked to make a final submission and present a BAFO. This is the moment you have been working toward when the proposal was initially submitted. It is now time for the final fine-tuning, incorporating everything gleaned from the customer's questions, hints in the trade press, and competitive intelligence. You do not necessarily have to reduce your price, that's not the reasoning behind a BAFO. Rather, think of it as your opportunity to modify your proposal to match any changes the customer has slipped into the final solicitation, so that you are bidding on the package they want to buy today, not the one they were seeking six months ago. Of course, you may be able to trim the price, a bit, as well.

Phase 6
Celebrate the win in an appropriately decorous fashion. Pass overall responsibility to the program manager and manufacturing. Have someone outside the team create a "lessons learned" report.

Congratulations! You've won the competition and soon enough—after whatever negotiations are deemed necessary—you will have a contract and have to start delivering on your promises. You will, of course, have continuing contact with the customer and regular performance reviews. The DoD issues a formal "Contractor Performance Assessment Report" (CPAR) beginning six months after the award and every 12 months thereafter until the job is done. International customers may have their own review process. These review processes and reports help keep you on target and help to identify and fix problems. Some contracts come with award incentives, rewards for above average performance. These look good on the record and can be used in your favor in the next proposal, plugged into the section on "Past Performance."

Very soon after every win or loss (not determined by size, but by the importance to the corporation) and before memories begin to fade or files disappear into some limestone cave, bring in an outsider—outside of the program—to compile lessons learned. You will find great support with large programs, especially in a loss, when everyone needs to know what happened. This may not be as easy on smaller programs, where funds are cut off as soon as the proposal is out the door, when the proposal team drifts off to other projects and their time is being charged to a different account.

As with running a WSSC, the case for using an outsider is clear, especially on a loss. A team member would be more likely to come up with CYA excuses: the customer always planned to go with our competitor, or somebody was bought off. Even if some of that may be true, it's not what you want to hear. You want someone to strip away the bullshit and uncover mistakes so you can avoid repeating them later. As I used to say to the team, "There is no education in the second kick of a mule."

Here's one useful lesson learned. Over a couple of competitions, whether or not we had won or lost, we saw that one company was underbidding us by a remarkably consistent percentage. Next time we went head-to-head, knowing that they were likely to be just that much cheaper, we took a hard look at our margins, looked for places to cut, and developed more powerful arguments as to why our products would cost more and were worth it.

You might also run this process on your Process. Create a score card and give yourself a grade. Did you have adequate resources—people, funding, facilities? Ask your team members and senior executives what are we doing that impedes success? What could we be doing to make things better, more effective? Were decisions timely? How did you do on market assessment? Competitive assessment? Did you have sufficient executive support? Were bids realistic and competitive? Did you really understand the customer's MIRS?

Phase 7
Keep it sold!

It seems like something that doesn't need to be said—but needs to be said. You won the bid, the ceremonial deliveries are past, the first groups of eager young pilots have been trained, and the nation that bought your airplanes is better protected against enemies real and imagined. But your job isn't finished.

Now come the nagging uncertainties. Perhaps there are unresolved issues. Something was promised that didn't happen. Perhaps it's buyer's remorse—they paid too much, or they didn't give your competitors a fair shot. With foreign customers, cultural differences begin to rub sore spots in your relationship. What you believed to be good old American willingness to help the less technologically advantaged they see as classic Yankee arrogance. Closer to home, natural and healthy rivalries between, say, the Air Force and the Navy are aggravated by sniping from their supporters (and your detractors) in Congress.

Just as in a marriage, even a very good marriage, unless caught in an early stage, simple irritation or mild depression may soon enough lead to open hostility. You were buddies. Last year, it was like a glorious fall, brilliant with a snap in the air. This year, winter has set in.

In every victory may be planted seeds of defeat. You have sold and delivered an amazing product but your job doesn't end with delivery of the first item, or even the last. This is where too many companies fall short, when total responsibility shifts from business development to program management. The customer doesn't know from "program management." Program management didn't sell them the airplanes.

Yes, you probably have a local office to provide support and assistance, but they may be seen as intermediaries. Not to disparage or belittle, but to be brutally frank—they may just be leave-behinds, the human equivalent of the brochure you might drop on a customer's desk, with very limited authority. Yes, you have tech reps standing by, on location, to flush out and fix transient problems, but do your tech reps get frustrated when a fastener is too tight or a circuit tester dies, and say something like, "This piece of shit...." Is your customer standing nearby? I don't mean your *customer*, but someone who works for him. Perhaps a junior maintenance tech, assigned to stick with your man and learn, who may be unfamiliar with American slang. If your *customer* hears that your own representative is unhappy with your product, what is the customer to think especially when your product does in fact begin to exhibit normal growing pains?

A fuel gauge stops working. A power supply demonstrates a life cycle about half of what had been promised. And, perhaps, your competitor did not just crawl off to his cave to lick wounds. Perhaps he can't upset *this* deal, but there will be others in the future. As an industry insider, he will soon enough know what issues may come along, and know them before your client. Your competitor wants to be the first to be the bearer of bad tidings.

Your marketing team is the face of the company, and from the years you spent cultivating your customers, you should know them well enough to pick up the early signs of a relationship going sour: a certain emotional distance, perhaps a disinterest in discussing future projects. But you can only detect these things by maintaining the relationship. Let program managers manage the program; you and your team need to be out in front, continuing to represent your company to the world. Drop by your customers every once in a while, prepared not to sell or to lecture but to assess attitudes, listen, take notes, and take action.

There's a problem with some obscure but vital part? Don't lay the blame on a supplier: your customer doesn't care, his contract is with you. He wants his airplanes ready to fly at a moment's notice. For known problems that have not yet surfaced in the customer's fleet—the program managers had better keep you in the loop. Perhaps you and the program manager (who does, after all, have an important role to play) should get on an airplane and do a face-to-face with the customer well before he hears about the problem from some outside source. For a *really* serious issue, take your chairman along. The customer should be properly impressed by the commitment thus made visible and you and the program manager most likely will then get to fly in a company plane. This makes for a less stressful trip and leaves you fresh for the client meeting.

Phase 8
Recycle your marketing team, go back to Phase 1, and look for another project.
You get the idea . . .

Epilogue

On August 1, 1997, McDonnell Douglas and Boeing, the leading builder of military aircraft and the leader in commercial airplanes, became one, creating the world's largest aerospace company. I participated in the merger task force and served as chair of the Business Development team. We spent six months working through the details. However, for tax and other financial reasons, both companies soon realized that the deal that was planned as a merger would work to the best benefit of the shareholders if Boeing would buy McDonnell Douglas.

It was, for some of us, the passing of an era as the names of two aerospace pioneers disappeared from the marketplace, but not from our memories. I believe there are very few employees who can say they actually worked for all four of the McDonnells—company founder, James (Mr. Mac), his eldest son, Jim (for whom I was a special assistant in my first corporate-level posting), Mr. Mac's nephew, Sanford (chairman and CEO, 1980–1988), and son, John (CEO from 1988–1994 and chairman from 1988 until the sale, when he became a member of the Boeing board of directors). I treasure the experience and know that the lessons learned from each played a major part in my personal and professional growth. The McDonnells shared key traits: they were exceptionally bright, completely fair, and displayed a natural humility that is rare in an era where the ego of an industry leader is often dwarfed only by the size of his paycheck.

Mr. Mac taught me the importance of "detail": you don't settle on an answer until you have exhausted yourself and your team questioning assumptions, verifying facts, and understanding the second order consequences of your action. From Jim I learned about the customer, the international environment, cultural sensitivity, and financial discipline. Sandy was great with people. He had a "program manager's" style of running the company and of valuing individuals. He stressed the importance of honest and ethical actions and believed that each person had a great talent and it was a responsibility of management to find out what it was and develop it to the fullest.

John was the business man. He understood finance, risk, organizational imperatives, the need to set goals that were not easily met but made everyone work harder, and the importance of having good products, global markets, and a dedicated work ethic. He was truly attuned to business development and marketing (he knew that relegating either to the corporate basement, so to speak, was like trying to grow grass on a sidewalk. It will take root only in the cracks, and then not very well). John willingly traveled the world in support of our efforts. In the three-year period before his retirement, he visited some thirty nations—many, more than once—to extol the excellence of our products (and also to endure bad meals, lost sleep, and countless less-than-fascinating tours of local "attractions" all to help our company).

As for me—as of the date of the sale—I swapped a full-time job for retirement and a contract as an on-call consultant for Boeing with the option to consult with other companies as well. It keeps me busy but has also allowed me the time to assemble my thoughts and pass them along in this book, which, by now, I presume you will have read.

How may I best summarize and give you a take-away for the future? I can reprise what by now you certainly recognize as key themes. I believe that the sale is the most important factor in business. *Most important.* What about "shareholder value," the catch phrase of the day? No sales, no value. Start at the beginning. You can have superb engineering and great products or great ideas but if you can't get them to market and sell them to a willing buyer, you're dead. Getting them to market and selling them is the job of marketing. As Jim Beveridge postulated back on page 19, engineers make lousy marketers, manufacturers should manufacture and program managers should manage programs. Marketing and selling should be handled by the professionals who know how to do this stuff, and they should be seated in the corporate councils as co-equals with the folks who create the product. Planning for the sales should start long before there is a product to sell, and if properly integrated, will help shape the product to the market.

Marketers must know the product and know the customer. They should initiate, maintain, cultivate, expand, and direct all contact with the customer. Marketing requires people with excellent communication skills, independence, and success-oriented leadership, maturity, judgment, fortitude, and personal integrity. They must be self-motivated and self-reliant, strong enough to stand up to senior executives and program managers, and to command respect from all. They must also follow a well-developed, disciplined process that will lead them from the possible to the probable.

In keeping with the punning title of the book, let me offer, as final advice, the following "Gunn Points." You might want to copy the list and tape it to your bathroom mirror; this should give you a good motivational start on the day.

> **Point One**: Do everything that could or should be done to win.
> **Point Two**: Always put yourself into the competitor's and decision maker's shoes.
> **Point Three**: Line up all the forces involved to support your case and neutralize those that don't.
> **Point Four**: Research every critical factor involved in the sale carefully, unemotionally, and responsibly.
> **Point Five**: Leave nothing to chance. Make your plan consistent, flexible, upgradable, motivational, thorough, and detailed.
> **Point Six**: Make all your presentations simple, fundamental, and brief.
> **Point Seven**: For each significant customer know what he thinks, how he thinks, what sort of information is best supplied to him, understand his past and present, and understand where he is going and what your proposal means to his probable future.
> **Point Eight**: Effective marketing requires the customer to hear, see, understand, be motivated by, and retain key messages of a campaign.
> **Point Nine**: Deliver your message to each person affecting the complex procurement and not just those who give you the best reception.

Point Ten: Assess the biases and inaccuracies of your proposal and the customer's request.

Point Eleven: Carefully interpret facts; never jump to conclusions—learn to listen.

Point Twelve: Draft and present a proposal with a competitive price from which you are the logical winner.

Point Thirteen: Be ready to "stop" unproductive initiatives and redirect your energies to projects likely to be more profitable.

Point Fourteen: "Selling" does not end when the product is delivered.

Point Fifteen: YCSASOYA. (Go back to page 21.)

Index

A6 Intruder, 156, 157
A-12 Avenger II, 22–23, 25, 26, 28, 156, 157
A-64A attack helicopters, 26
A-400M, 101
A. T. Kearney (management consultants), 29, 30
A300-600s, 139
Addabbo, Joe, 69
Advanced Manufacturing Research Center, University of Sheffield, 106–7
Advanced Medium STOL (Short-takeoff and Landing) Transport (AMST), 109
advanced tactical fighter, 25
aerospace marketing, 9, 11–35; Best and Final Offer (BAFO), 15; how governments buy airplanes, 13–14; inducements (bribes) to make a sale, 15–17; new model production, 13–14; prototypes, 15; request for proposals (RFPs), 15; research and development (R&D) funding, 15; timeline for building military airplanes, 12; traditional sales model, 11; twin-engine versus single-engine airplanes, 14. *See also* international marketplace; sales
aerospace programs, 8
aerospace trade press, 34, 81, 86–87
Aerospatiale, 146, 147
African Americans, 47
aging employees, 48
AH-64A Apache Attack Helicopter, 13, 26, 66, 82–83; Netherlands, 153; and UK, 128–29; women's pantyhose, 83
AH-64D "Longbow" Apache, 13, 26, 153
AIPAC. *See* American Israel Public Affairs committee
Air Force Advanced Tactical Fighter program (ATF), 26–28
air shows, 34, 79–83; aerospace trade press, 81; entertainment versus marketing, 79; and espionage, 80; Navy's Blue Angels, 79; Paris Air Show, 79, 82, 126; public and private demonstration flights, 81; reasons for exhibiting, 82; recruiting pilots, 79; relationship building at, 80

INDEX

air-to-ground attack planes, 14
Airbus, 139
Airbus Military, 101
airplanes, at McDonnell Douglas, 12–13
Al Yamamah (The Dove), 131, 133, 134
American Israel Public Affairs committee (AIPAC), 130–31
American military aerospace, 11
AMP. *See* Avionics Modernization Plan
AMST. *See* Advanced Medium STOL (Short-takeoff and Landing) Transport
Anatomy of a Win: Aerospace Marketing, The (Beveridge), 18–20, 22, 25–26, 43, 182
Andrews Air Force Base, fly-off of fighter models at, 84–85
anti-ship missiles, 8, 9, 146, 149. *See also* Harpoon
Appropriations Committee (House), 8–9, 68
Appropriations Committee (Senate), 5, 6, 68
Arab world cultural issues, 62–63
Argentine-U.K. war in the Falklands, 146
Armed Services Committee, 68
Art of War (Sun Tzu), 42
Art of War, The (Machiavelli), 42
Ashcroft, John, 137
Astronautics Company (McDonnell Douglas), 8
ATF. *See* Air Force Advanced Tactical Fighter program
athletic cup, 3
attack helicopters, 14
Australia, F-18 versus F-16, 53

authority to make decision, 32, 34, 129
AV-8A, 154
AV-8A Harrier, 13
AV-8B Harrier II, 154, 159
AV-8B Harrier II Plus, 13, 154, 155
Avionics Modernization Plan (AMP), 158–59

B-1 bomber, 8
background checks, 51
BAE (British contractor), 154, 159
BAFO. *See* Best and Final Offer
Baker, Jim, 137
Base Realignment and Closure Commission (BRAC), 157–58
Bass, Mark, x
Bell Helicopter Textron, 128
Berlin Wall, 124
Berman, Howard, 130, 138
Best and Final Offer (BAFO), 15, 77
Beveridge, James M., *Anatomy of a Win: Aerospace Marketing, The*, 18–20, 22, 25–26, 43, 182
Bin Abdul Aziz, Fahd, 139
Bin Sultan, Bandar, 132
Blair, Tony, 100, 107, 134
"block" designations, 14
body language, 39
Boeing, 26, 27, 70, 106; and C-130, 159; and Harpoon, 150; McDonnell Douglas merger with, 157, 160, 181; and modified 747 freighter, 108, 109, 113
Boeing 737s, 139
Boeing 747s, 139
Boeing/Sikorsky, 25, 26, 128
Bold Stroke project, McDonnell Douglas, 159

INDEX

Bond, Kit, 137
boxing lessons, 3
BRAC. *See* Base Realignment and Closure Commission
bribes. *See* inducements (bribes) to make a sale, 15–17
British Aerospace, 127, 134
Bush, George H. W., 132–33, 136, 137–38
business meetings, 60
business plan, 1, 20
buyer's remorse, 1178

C-5, 88, 109, 111, 116
C-17 Globemaster II Long-Range Airlifter, 90, 159; Advanced Medium STOL (Short-takeoff and Landing) Transport (AMST), 109; case study, 108–16; Collier Trophy, 116; description of, 13; marketing challenges of, 31; Operation Allied Force (1999; in Yugoslavia, 116; personnel needed for, 88; United Kingdom sale, 100–101
C-130 cargo plane, 100, 109, 158–59
C-141, 109
C-X (Cargo Transport Aircraft-Experimental), 109, 110, 111
Caldwell, Jim, x, 133, 136
Calloway, Jim, 6
candidate characteristics, for team in international business development, 37–38, 183
candidate files of consultants and agents, 57
careers in international business development, 36–37
Carter, Jimmy, 70, 109

case studies, 108–59; C-17 Globemaster II Long-Range Airlifter, 108–16; F-18 "upgrade," 156–57; Finland F-18 sales, 124–28; Harpoon, 146–51; Harrier, 154–55; Joint Direct Attack Munitions (JDAM), 142–46, 149; Korean F-18 adventure, 117–20; Malaysia, 151–53; Military Aerospace Support Division, 157–59; Netherlands, 153–54; partnerships, 128–30, 154–55; Saudia Airlines, 139–41; Spain F-18 sales, 120–22; Switzerland F-18 sales, 122–23; Thailand, 155–56; "U.S. Jobs Now," 130–39
Cassidy, Gerry, 70
Central Intelligence Agency (CIA), 7, 79, 93
Chadwick, Chris, x
characteristics of candidates, for team in international business development, 37–38, 183
Charity, Sister Mary, 164
Chautauqua circuit, 2
checkpoints (gates), 161
Cheney, Dick, 136
Clay, Cassius (Muhammad Ali), 3
Clinton, Bill, 136, 139
clothes, 60
Coggins, Mike, x
co-production and unions, 106
Cold War, end of, 135
college transcripts, 38
Collier Trophy, 116
Comanche RAH-66, 26
commercial sales, 30
Committee on Commerce, Science and Transportation (Senate), 69

common sense and courtesy, 65
company organization, 44
competition, 83–90, 171; analysis of, 20, 29; "fair advantage," 86; fly-off of fighter models at Andrews Air Force Base, 84–85; foreign, 17; "full and open, best value," 94–95; "gamemanship" in, 86–87; international, 87–90; knowing the, 11; nationalized or government-subsidized companies, 78, 89; new products of, 2, 13; rules (or no rules) of the game, 83–84; writing periodic report for aerospace industry contractors, 85–86
compromising situations, and espionage, 91–92
confidentiality agreements, 145
conflicts of interest, 51
Congress, 66–71; Foreign Corrupt Practices Act (FCPA), 16–17; gatekeepers, 67; keeping members informed, 7–68; local district offices, 67; McDonnell Douglas Political Action Committee (MDC-PAC), 66–67; personal relationships, 69–70; PR consultants, 67–68, 69; selling to the government, 66; as source of money, 66; and special interest groups, 67; "U.S. Jobs Now," 130–39; visibility of members of, 66
consultants and agents, 50–59; background checks, 51; conflicts of interest, 51; contracts, 58; and cultural issues, 58; file of candidates, 57; international operations, 51; lessons learned in hiring, 58–59; reasons for hiring, 51–52; retired military officers, 52–53; selection of, 56–58
"Contractor Performance Assessment Report" (CPAR), Department of Defense (DoD), 177
contracts, 177; consultants and agents, 58; with Department of Defense (DoD), 15, 90
"Convention on Combating Bribery of Foreign Public Officials in International Business Transactions," Organization of Economic Cooperation and Development (OECD), 17
cost plus contracts, 144
counter-trade, and industrial participation (offset), 103
country analysis, 78–79
courtesy and common sense, 65
"courtesy calls" after a sales loss, 101–2
Craighead, Bob, x
cultural issues, 45, 59–66, 178; in Arab world, 62–63; business meetings, 60; clothes, 60; common sense and courtesy, 65; and consultants, 58; examples of missteps, 63–65; knowledge of and sensitivity to, 59; meals and eating, 61; status-evaluation, 60; surveillance, 59–60
culture, 45
customers, 1; knowing about, 31, 183; listening to, 29; specific requests of, 24
customer's problem, 31
Czech Republic, 105

INDEX

Daniel, Dan, 110–11
Dassault, Serge, 92, 129–30, 134
Davis, Darryl, x
De Bono, Edward, "Thinking Hats," 120–21
Decision Mapping, Hodapp, Richard, 25
decisions, ultimate authority to make, 32, 34, 129
Delft University, 106
Democratic Party, 4
demonstration flights, 81
Denmark, Harpoon II, 150
Department of Defense (DoD): and Appropriations Committee, 6; budgeting in, 76; "Contractor Performance Assessment Report" (CPAR), 177; contracts with, 15, 90; Future Years Defense Plan (FYDP), 76; and Gunn, 6; "Specifications and Standards—A New Way of Doing Business," 143–44
Desena, Louis, 69
design campaign and get approval to make a bid (Gate 3) in the Process, 170–72
develop proposal and get approval to submit it (Gate 4) in the Process, 172–75
develop winning strategy and get approval (Gate 2) in the Process, 167–70
Dillow, Charlie, 144
Dorrenbacher, Jim, 25, 29
Douglas, 21, 30, 31, 65
downsizing, 49–50
Dubai Airshow (1991), 132, 134
Duberstein, Ken, 70

EA-6, 157
Eagleton, Tom, 4
education of Gunn, 2, 4
egos and sales, 75
electronic listening devices, 93
Eliat (Israeli destroyer), 146
employment contracts, 47–48
English School, Finland, 125
Ervin, Sam, 6
espionage, 90–93; and air shows, 80; compromising situations, 91–92; electronic listening devices, 93; hotel rooms, 91, 92–93; laptops, 91; surveillance, 91; telephones, 92; United States embassy in Moscow, 93; wristwatch incident, 92
Eurocopter, 128, 153
European Future Large Aircraft (FLA), 100
Executive Reviews (gates), 161
Exocet (French anti-ship missile), 146, 147, 148, 149, 150, 151
export license, 89

F/A-18, 12
F/A-18E/F Super Hornet, 157
F-4 Phantom, 78, 102, 125–26
F-14, 85, 87–88, 157
F-15, 41, 82, 85, 102, 159; Saudi Arabia sales, 130–39
F-15A Eagle, 12, 14
F-15E Strike Eagle, 12, 14, 120, 138
F-15S, 139
F-16, 25, 54, 102, 138; Australia, 53; Finland, 124; Harpoon missile on, 89; Malaysia, 152; NATO fighter program, 121; Thailand, 155–56;

F-16 (*continued*)
  Turkey, 103; versus F-18 in Korea, 117–20
F-18 Hornet, 12, 21, 25, 26, 27, 41, 159; in air shows, 79; Australia, 53, 103, 122; Canada, 103, 122; case study, 117–20; Finland sales, 124–28; Iran, 87–88; Korea, 54, 117–20; Malaysia, 152; Spain sales, 120–22; Switzerland sales, 54, 122–23; "upgrade," 156–57
F-18C/D, 156
F-20, 123
F-22, 14, 27
F-35, 14
F-117 stealth fighter, 18
F-151, 138–39
"fair advantage," 86
Farnborough International Air Show (London), 55
FCPA. *See* Foreign Corrupt Practices Act
Federal Bureau of Investigation (FBI), 55
Federal Trade Commission, 5
Feren, John, *x*
field offices, 78–89, 179; country analysis, 78–79; good will, 78
Finland; English School, 125; F-18 sales, 124–28
Finnair, 124
fire control radar, 13
firing employees, 47–48
Fisher, Martin, *x*, 147, 149
fly-off of fighter models at Andrews Air Force Base, 84–85
focus group sessions, 1344
Fokker (Dutch company), 153–54

foreign competition, 17
Foreign Corrupt Practices Act (FCPA), 16–17, 57, 58; and industrial participation (offset), 103; release of technology, 87
Foreign Military Sales (FMS) program, 148
forms of government, 76–77
Fortas, Abe, Supreme Court, 6
4E formula (Jack Welch), 42
France, 87, 89, 93; and McDonnell Douglas, 129–30
funding schemes, request for proposal (RFP), 95–96
Future Years Defense Plan (FYDP), Department of Defense (DoD), 76

"gamemanship" in competition, 86–87
GAO. *See* Government Accountability Office
gatekeepers in Congress, 67
gates (Executive Reviews), 161
GEC-Marconi, 128
General Dynamics, 23, 25, 26, 27, 54; A-12 Avenger II, 156; carrying Harpoon missile, 89; and Finland, 124; Korean F-16 adventure, 117–20; offset in Greece, 104; offset in Turkey, 103; Taiwan, 137; trainer market, 106
General Electric, 27, 42, 104, 133, 152
General Services Administration (GSA), 7
Gephardt, Dick, 136
Gingrich, Al, *x*, 22–23
GKN, 129
Global Strike Division, 41
goals, and strategies, 21

INDEX

Golden Gloves, 3
Goldwater, Barry, 110
good will, 78
Government Accountability Office (GAO), 30
government-military market, 30
Government Operations Committee, 5
Governments: buying airplanes, 13–15; forms of, 76–77; selling to, 66
GPS-guided steering systems, 142, 149
Grace, Bob, x
graphics, 174
grass-roots marketing, 130
gravity bombs, 142
"grease payments," 16
Greece, 103–104
Grumman, 26, 27, 85, 87–88, 156, 157
Gulf War (first), 116, 131; Army Apache AH-64 helicopters in, 66, 82–83
Gunn, Donald, 2, 4
Gunn, Kate, 4, 70, 71, 115
Gunn, Megan, 71
"Gunn Points" (summary), 182–84
Gunn, Thomas: boxing lessons, 3; and Department of Defense, 6; education of, 2, 4; establishes New Business Activity, 30; first sale, 9; "Gunn Points" (summary), 182–84; hired by McDonnell Douglas, 7–8; law school, 4; lessons learned, 3–4; marketing manager, Astronautics (1975), 8; marriage to Kate, 4; staff vice president McDonnell Douglas Washington operations (1983), 17; politics, 4; president of Helicopter Company (1990), 25, 48; retirement and contract as on-call consultant for Boeing, 182; security clearances of, 6, 8; vice president of marketing (1986), 20; White House job offer, 70; working for John McClellan, 5–7; working for Symington's law firm, 5; youth and family life of, 2–4

handicapped employees, 48
Harpoon, 142, 143, 146–51; anti-ship missile program, 8, 9; Harpoon Block 1G, 148, 150; Harpoon Block III, 151; Harpoon II, 150; "Harpoon Myths and Facts," 148; international sales, 147; land attack missiles, 149, 150; Malaysia buying, 152
Harrier, 154–155
Harris, Brayton, xi
Helicopter Company, McDonnell Douglas, 25, 48, 83, 128
Hennelly, Loretto (Mrs. Donald Gunn), 2, 5
Hennelly, Mark, 5
Hibbard, George, x
Hodapp, Richard, Decision Mapping, 25
hotel rooms, and espionage, 91, 92–93
Hughes Aircraft Company, 19, 133, 144, 152
Hughes Helicopters, 13
hypothetical questions in job interviews, 39–40

identify and prioritize opportunities (Gate 1) in the Process, 162–67
identity dysfunction, 43–44
"ilities," in sales, 78
"Ill Wind," 54–55

INDEX

improvised explosive devices (IEDs), 14
incentive compensation (IC) program, 46–47, 145–46
inducements (bribes) to make a sale, 15–17
industrial espionage. See espionage
industrial participation (offset), 11–12, 65, 86, 102–7; Advanced Manufacturing Research Center, University of Sheffield, 106–7; benefits of, 105, 107; co-production and unions, 106; and counter-trade, 103; defined, 102–3; and FCPA, 103; internship programs, 106; in less-developed countries, 105; in Saudi Arabia, 105; with, Spain, 104–5
influencers, internal and outside, 32
Integrated Product Teams (IPT), 144–45, 167
intellectual property, 168
internal and outside influencers, 32
international business development career, 36–37
international competition, 87–90
international customers, 73–74
international marketplace: country specialists, 30; Harpoon, 147; how governments buy airplanes, 14–15; inducements (bribes) to make a sale, 15–17; rules governing sales, 16. See also aerospace marketing
International Space Station, 116
international versus domestic sales, 73
internship programs, industrial participation (offset), 106
Iran, 87–88, 134
Iranian Revolution, 85, 134
Iraq, 136, 140

Iraq-Iran tanker war (1980–1988), 147
Israel: American Israel Public Affairs committee (AIPAC), 130–31; and Middle East arms sales, 133, 134; and Saudi Arabia, 130, 139
Italy, 154
ITP. See Integrated Product Teams

jargon, 45
JAS 39 Gripen, 123
Javits, Jacob, 70
JDAM. See Joint Direct Attack Munitions
Jeffords, James M., 138
job candidates, sources for, 40
job interviews, 39–40
Johnson, Lyndon, 6
Joint Direct Attack Munitions (JDAM), 142–46, 149
Joint Rapid Reaction Force (JRRF), 100
Jordan, Hamilton, 70–71
JRRF. See Joint Rapid Reaction Force
Judiciary Subcommittee on Criminal Laws and Procedures, 5

KC-10 dual role tanker-airlifters, 109, 110, 116
KC-135, 158
Kelly Air Force Base, San Antonio, 158
Kissinger, Henry, 152
Korean F-18 adventure, 117–120
Korean Fighter Program, 21–22, 25, 26, 117
Krause, Steve, x
Kronenberg, Mark, x
Kuwait, 82

L-3 Communications, 159
L-90 Redigo, 127
L1011, 139, 140
labor unions, 135; and co-production, 106
land attack missiles, 149, 150
laptops, and espionage, 91
Lavi (Israel), 123
law school, Gunn, 4
layoffs at McDonnell Douglas, 132
leaders for team, 41–43
leadership styles, 42–43
Lee Kuan Yew, 64
lessons learned, 165; after a loss "courtesy calls," 102; from all four of the McDonnells, 181–82; in competition, 85; in Golden Gloves incident, 3–4, 83; in hiring consultants, 58–59; in industrial participation, 104; report on, 177
LHX program, 25, 26–27
Lloyd's of London, 147, 148
local district congressional offices, 67
Lockheed, 18, 26, 27, 88, 90; and C-5B line, 109, 111, 113, 114
Lockheed Martin, 142, 158
Longbow Hellfire air-to-ground missile, 13
lost competitive sales, McDonnell Douglas, 150–51

MacArthur, Douglas, 4
Machiavelli, Niccolo, *Art of War, The*, 42
maintain customer relationships in Phase 7 of the Process, 178–80
Major, John, 134
Malawi, Cindy, x
Malaysia, 151–53

Malvern, Don, 19
Mapping Alliance, The, 15
market: government-military, 30; studying the, 1, 29
marketing plan, 167
Martin Marietta, 133
McClellan, John, Gunn working for, 5–7
McDonnell Douglas, 27; Boeing merger with, 157, 160, 181; Bold Stroke project, 159; C-17 case study, 108–16; Gunn hired by, 7–8; Helicopter Company, 25, 48, 83, 128; layoffs at, 132; lost competitive sales, 150–51; New Business Activity, 30–31; New Business Center, 23–25; "Pacific Dialogue," 152; Plans and Strategy Reviews, 20; R&D proposals, 23; and stealth technology, 18; types of airplanes, 12–13; Washington operations, 17–20; "YCSASOYA," 21, 184
McDonnell Douglas Political Action Committee (MDC-PAC), 66–67
McDonnell, James S. ("Mr. Mac"), 7–8, 13, 17, 73, 118, 181
McDonnell, Jim (son of James S.), 17, 181–82
McDonnell, John, 25, 27, 48, 118, 181, 182; and C-17 program, 114, 115–116; letter to Bush (GHW), 136; Malaysian visit, 152; Saudia's 50th anniversary, 141; Singapore, 64; toast in China, 61–62
McDonnell, Sanford, 181, 182
MD-11 (three-engine, long-range commercial airliner), 13, 64, 139, 140, 141, 152

MDC-PAC. *See* McDonnell Douglas Political Action Committee
meals and eating, 61
media, working the, 34
media training sessions, 46
Melroy, Pamela, 116
Merritt, Larry, x
Middle East: arms sales to, 132–34; industrial participation in, 105
MiG-29, 152
military aerospace: alliances with private industry, 158; American, 11
Military Aerospace Support Division, 157–59
military airplanes: components of, 13; timeline for building, 12
military budgets, 76–77
Mirage 2000 and 2000-5, 123
mission, 1, 43–44
mission statement, 21
money sources, Congress as, 66
money trail, sales, 75
Moscow, United States embassy in, 93
Most Important Requirements (MIRS), 32–33, 153, 178

national security, 90
National Security Agency (NSA), 7
nationalized or government-subsidized companies, 78, 89
Naval Air Depot at Cherry Point, NC, 155
Navy's Blue Angels, 79
Netherlands, 106, 153–54
Networking, 69
New Business Activity, 30–31; changed name to Program Development, then to Business Development, 72; objectivity, 34–35; 12 step process, 30–34
New Business Center, 23–25; formal business development process, 23–25; Red Teams, 24–25
new model production, 13–14
new products of the competition, 2, 13
Nixon, Richard M., 84
non-compete clause, 48
non-salesmen, 41
Northrop, 26, 27, 123, 128, 133, 152

objectivity in sales and marketing, 34
OECD. *See* Organization of Economic Cooperation and Development
Office of Management and the Budget (OMB), 76, 109
office space, 44
offset. *See* industrial participation
Olsen, Gerry, x
OMB. *See* Office of Management and the Budget
Omnibus Crime Control and Safe Streets Acts of 1968 and 1970, 6
Operation Allied Force (1999) in Yugoslavia, 116
oral presentations, 176; request for proposal (RFP), 97, 98
organization of company, 44
Organization of Economic Cooperation and Development (OECD), "Convention on Combating Bribery of Foreign Public Officials in International Business Transactions," 17
Organized Crime Control Act, 6
out placement assistance, 48

outsider teams, 174, 77
outsourcing, 56; within company, 56

"Pacific Dialogue," 152
Paisley, Melvin, 54–55
pantyhose, 83
Paris Air Show, 79, 82, 126
parliamentary systems, and sales, 74, 76
partnership case studies, 128–30, 154–55
pass responsibility to the program manager in Phase 6 of the Process, 177–78
payment plan in sales, 76
Penguin (Norwegian anti-ship missile), 146
Pentagon, 8
personal relationships in Congress, 69–70
pilot recruitment, 79
Plans and Strategy Reviews, McDonnell Douglas, 20
plant visits, 34
poker sessions, 44
political issues, 34, 35
post-Vietnam "Peace Dividend," 11
Powell, Colin, 136
Pratt & Whitney, 27, 28
pricing, 77–78, 170; time sensitive nature of, 78; "win price," 77
private industry, alliances with military, 158
Process, 160–80; Phase 1, identify and prioritize opportunities (Gate 1), 162–67; Phase 2, develop winning strategy and get approval (Gate 2), 167–70; Phase 3, design campaign and get approval to make a bid (Gate 3), 170–72; Phase 4, develop proposal and get approval to submit it (Gate 4), 172–75; Phase 5, submit Best and Final Offer (BAFO), 175–77; Phase 6, pass responsibility to the program manager, 177–78; Phase 7, maintain customer relationships, 178–80; Phase 8, recycle marketing team, go back to Phase 1, 180; scoring and grading the Process, 178
program managers, 180
progress report cards with color coded grades, 145, 146
prototypes, 15
public relations, 44–46
public relations consultants, 67–68, 69

Rabin, Yitzhak, 136
Radar: fire control, 13; and stealth technology, 18
Raymond, Chris, x
RBS-15 (Swedish anti-ship missile), 146, 150
Reagan defense build-up, 11, 69, 110
real-world example, request for proposal (RFP), 100–101
recruiting pilots, 79
recycle marketing team, go back to Phase 1 in Phase 8 of the Process, 180
Red Teams, 24–25, 168, 170, 171
Rehn, Elizabeth, 126–127
relationship building at air shows, 80
reports for aerospace industry contractors, 85–86
request for proposal (RFP), 15, 24, 29, 93–102, 170; before RFP is

request for proposal (*continued*)
published, 98–99; components of, 94–98; defining set of requirements, 98; "full and open, best value" competition, 94–95; funding schemes, 95–96; issuing, 172; oral presentations, 97, 98; real-world example, 100–101; unannounced customer needs, 99–100; U.S. Army's for first airplane in 1907, 93–94

research and development (R&D), 1, 151; funding for, 15; proposals by McDonnell Douglas, 23

Restelli, Jim, *x*, 157, 158

resumes, 38

retired military flag officers, as consultants, 52–53

RFP. *See* request for proposal

Rice, Donald, 27, 28

Rockwell, 26, 27

Rolls Royce engines, 129, 155

Roman, George, *x*

Rubik's Cube, 41

rules governing sales, 16

"rules of the game," 4, 83–84

rules (or no rules) of the game in competition, 83–84

Russian airplanes, 127

Russian Antonev aircraft, 100

Saddam Hussein, 82

sales, 72–78; after a loss "courtesy calls," 101–2; Best and Final Offer (BAFO), 77; feeding egos, 75; forms of government, 76–77; grass-roots marketing, 130; "ilities," 78; international customers, 73–74; international versus domestic, 73; military budgets, 76–77; money trail, 75; as most important factor in business, 182; nationalized or government-subsidized companies, 78; and parliamentary systems, 74, 76; payment plan, 76; pricing ("win price"), 77–78; rules governing sales, 16; sales calls like job interviews, 74. *See also* aerospace marketing; case studies; international marketplace

Saudi Air Force, 103

Saudi Arabia, 87; Al Yamamah (The Dove), 131, 133, 134; industrial participation in, 105; McDonnell-Douglas selling to, 130–39

Saudia Airlines, 139–141

Science and Technology Committee (House), 69

scoring and grading the Process, 178

secret development projects, 13

Securities and Exchange Commission, 45

security clearances of Gunn, 6, 8

selling missiles and airplanes. *See* aerospace marketing; sales

Shah of Iran, 84–85

shareholder value, 182

Sheffield (British destroyer), 146

Shorts Starstreak missiles, 129

Six Days War (1967), 146

SLAM. *See* Standoff Land Attack Missile

Soviet Union, disintegration of, 127

Spain, 154; F-18 sales, 120–22; industrial participation with, 104–5

special interest groups, 67

"Specifications and Standards—A New Way of Doing Business," Department of Defense (DoD), 143–44
St. Louis Board of Aldermen, 4
Standoff Land Attack Missile (SLAM), 149, 150
Stark, 147
status-evaluation, in cultural issues, 60
stealth technology, 8, 18
storyboards, 23–24, 25
strategy, and goals, 21
Stroupe, Regina, x
Styx (Soviet anti-ship missile), 146
submit Best and Final Offer (BAFO) in Phase 5 of the Process, 175–77
suicide bombers, 14
Sukhoi Su-27/35, 155
Sullivan, Mark, x
Sun Tzu, *Art of War*, 42
Surveillance: in cultural issues, 59–60; espionage, 91
Switzerland F-18 sales, 122–23
SWOT analysis, 33
Symington, Stuart, 4, 5
Syria, 136
System Design and Development, 12

Tate, Dan, 70
Tavernetti, Len, x
team for international business development, 37–50; body language, 39; characteristics for candidates, 37–38, 183; college transcripts, 38; employment contracts, 47–48; firing, 47–48; hypothetical questions, 39–40; identity dysfunction, 43–44; incentive compensation (IC) program, 46–47; job interviews, 39–40; leaders for team, 41–43; leadership styles, 42–43; mission, 43–44; non-salesmen, 41; office space, 44; organization of company, 44; out placement assistance, 48; poker sessions, 44; public relations, 44–46; resumes, 38; sources of job candidates, 40
technology: advances in, 11; and the market, 1
Teledyne Ryan Aeronautical, 7
telephones, and espionage, 92
television interviews, 45–46
Texas Instruments, 148
Textron, 128
Thailand, 155–56
Thatcher, Margaret, 87, 131
"Thinking Hats" (De Bono), 120–21
Thompson, Fred D., 6
Thompson, Stu, x
threat matrix, 165
thrust vectoring, 28
timeline for building military airplanes, 12
Tinker Air Force Base, Oklahoma City, 158
Tomahawk missile (Navy), 144
traditional sales model, 11
Truman, Harry, 4
Turkey, 103
12-step process in New Business Activity, 30–34; 1) customer's problem, 31; 2) addressing problem, 31–32; 3) ultimate authority to make decision, 32, 34, 129; 4) internal and outside influencers, 32; 5) Most Important Requirements

12-step process in New Business Activity (*continued*) (MIRS), 32, 153, 178; 6) prioritizing MIRS, 32–33; 7) issues, 33; 8) resolving issues, 33; 9) drafting the offer, 33; 10) SWOT analysis, 33; 11) writing up strategy, 33–34; 12) telling the customer about offer, 34
twin-engine versus single-engine airplanes, 14

U.K. *See* United Kingdom
ultimate authority to make decision, 32, 34, 129
unannounced customer needs, request for proposal (RFP), 99–100
United Embassy in Moscow, 93
United Kingdom, 87, 89, 100–101; Al Yamamah (The Dove), 131, 133, 134; attack helicopters, 128–29; C-17 sale to, 100–101; University of Sheffield, 107
United States embassy in Moscow, 93
United Technologies, 133
University of Sheffield, England, 107
U.S. Army's for first airplane in 1907, request for proposal (RFP), 93–94
"U.S. Jobs Now," 130–39

Van Gels, John, *x*
V/STOL (vertical/short take off and landing) attack airplane, 13

Vietnam War, 6
visibility of members of Congress, 66
vision statement, 21

war room, 169, 170, 173
warrantees and guarantees, 171
Washington operations, McDonnell Douglas, 17–20
Watergate Committee hearings, 6
Weaver, Sister Claire Marie, 125
Weinstock, Rich, *x*
Welch, Jack, 152; 4E formula, 42
Westland (UK) and McDonnell Douglas, 128–29
Whiteford, Fred, *x*
Win Strategy Steering Committees (WSSC), 29–30, 146, 169, 170, 171
women's panty-hose, 83
World Court in The Hague, 4
Wright brothers, 94
wristwatch incident, espionage, 92
WSSC. *See* Win Strategy Steering Committees

YC-14, 109
YC-15, 109, 110
"YCSASOYA," 21, 184
YF-22, 27, 28
YF-23, 27, 28
Yugoslavia, Operation Allied Force (1999), 116

About the Author

Tom Gunn had a life-altering career change in 1975 when he went from an eight-year stint as staff lawyer with the U.S. Senate to a job in aerospace sales and marketing at McDonnell Douglas. He knew a lot about military appropriations and classified developments, but almost nothing about marketing. Over the next twenty-two years, however, Gunn and the team he assembled developed a process for strategic selling and marketing that delivered $250 billion in sales of military and commercial aircraft, missiles, space systems, and logistic support, against strong and at times cutthroat domestic and international competition.

Gunn served in a number of positions with McDonnell Douglas, including vice president, marketing; president of the Helicopter Company, and president of McDonnell Douglas International. He is now a consultant to Boeing. Gunn and his wife Kate make their home in St. Louis, Missouri.

The Naval Institute Press is the book-publishing arm of the U.S. Naval Institute, a private, nonprofit, membership society for sea service professionals and others who share an interest in naval and maritime affairs. Established in 1873 at the U.S. Naval Academy in Annapolis, Maryland, where its offices remain today, the Naval Institute has members worldwide.

Members of the Naval Institute support the education programs of the society and receive the influential monthly magazine *Proceedings* or the colorful bimonthly magazine *Naval History* and discounts on fine nautical prints and on ship and aircraft photos. They also have access to the transcripts of the Institute's Oral History Program and get discounted admission to any of the Institute-sponsored seminars offered around the country.

The Naval Institute's book-publishing program, begun in 1898 with basic guides to naval practices, has broadened its scope to include books of more general interest. Now the Naval Institute Press publishes about seventy titles each year, ranging from how-to books on boating and navigation to battle histories, biographies, ship and aircraft guides, and novels. Institute members receive significant discounts on the Press's more than eight hundred books in print.

Full-time students are eligible for special half-price membership rates. Life memberships are also available.

For a free catalog describing Naval Institute Press books currently available, and for further information about joining the U.S. Naval Institute, please write to:

Member Services
U.S. Naval Institute
291 Wood Road
Annapolis, MD 21402-5034
Telephone: (800) 233-8764
Fax: (410) 571-1703
Web address: www.usni.org